Research
Design
in the
Behavioral
Sciences

Multiple Regression Approach

Research Design in the Behavioral Sciences

Multiple Regression Approach

Francis J. Kelly
Donald L. Beggs and
Keith A. McNeil with
Tony Eichelberger and
Judy Lyon

SOUTHERN ILLINOIS UNIVERSITY PRESS
Carbondale and Edwardsville

FEFFER & SIMONS, INC.
London and Amsterdam

Library of Congress Catalogue Card Number 69–15324
Printed by offset lithography in the United States of America
Standard Book Number 8093–2957–3

To Dave Miles, who said, "Write it."
Bruce Amble, who needs it;
Earl Jennings, who should have written it;
Don Veldman, who does not believe it;
Tom Jordan, who does; and
Traditional Statisticians and Operant Conditioners, who'll never read it.

Preface

This book is concerned with the development and the construction of multivariable statistical models. The use of such statistical devices assumes that the behavioral science investigator is asking a multi-dimensional question. If one looks at the literature in the behavioral sciences (i.e., psychology, sociology, education, economics, etc.), it is apparent that many factors underlie human behavior. On the other hand, if one focuses upon any particular study, it is also apparent that many of the variables which are theoretically relevant to the investigation are randomized, and only one or two factors are examined in a particular study. It seems that we recognize complexities regarding behavior but fail to conduct studies which reflect these complexities.

What causes such discrepancies? It is the *belief* of the present authors that many investigators' research is limited because the statistical models to which these investigators have been exposed are limiting. A review of recent statistical textbooks in the behavioral sciences shows that these writers assume the high-speed digital computer has neither been invented nor widely distributed. Consistently one is confronted with a series of factorial designs, Type I, Type II models, etc., derived to ease computation with a desk calculator. Unfortunately, it seems that such presentations either confuse the potential researcher or impose such constraints upon his designs that he is forced to ask a limited research question.

The multiple regression analysis presented in this book is designed to prepare the research investigator to construct statistical models which will reflect his original research question rather than limiting that question. Regression analysis will be shown to be the generalized case of analysis of variance. These discussions shall be intimately related to a computer program so that the simple elegance of the generalized analysis of variance is not obscured and so that the investigator can circumvent the anachronistic desk calculator.

Plan of the Book

The first chapter presents a brief behavioral model which the authors find helpful in summarizing the factors relevant to behavior. This model is a quasi-mathematical model and is related in the discussion to a few possible linear regression models.

A brief overview of inferential statistics and sampling theory is the focus of Chapter II, providing the basis for the notion of variance and least squares procedures.

Chapter III includes a discussion of vector notation and vector operations. The straightforward discussion presents vectors as simple identities for variables, facilitating an understanding of the mathematical operations used to build linear models.

Analysis of research questions which use mutually exclusive categories in the predictor set is the central theme of Chapter IV. An illustration of a univariate F is presented using traditional and regression procedures. The problem includes an empirical case. A number of expansions and possible statistical models are presented for illustrative purposes. The assumptions underlying these types of models are stated, and several problems are presented for use in conjunction with Chapter V.

The use of one linear computer program is detailed in Chapter V. The particular program has gone through a number of evolutionary stages and, as presented, provides the research investigator with a wide, flexible range. The program uses a variable format and gives means, standard deviations, a square correlation matrix, regression weights, multiple R's, an F test, and level of probability. One other valuable feature of the program is that a wide number of data transformations can be completed in the process of data analysis. Test problems from Chapter IV are presented, and the Appendix includes computer print-outs for checking purposes. The program given in Chapter V is a mainline program which calls for a number of sub-routines. The sub-routines can be placed in disc or tape storage, which saves computer read-in time. The sub-routines are in Appendix B. Two sets of sub-routines are presented (1) using FORTRAN IV adapted to IBM 7040, and (2) using FORTRAN IV adapted to IBM 360.

A return to model building and stating testible questions is the central facet of Chapter VI. The uses of continuous predictor variables singly or combined with categorical data and other continuous variables are the problem types discussed. Further assumptions regarding the model using continuous data are specified.

Chapter VI also presents non-rectilinear models that can be cast and which will reduce the number of assumptions which the investigator must tolerate. Curved, cubed, etc., functions are discussed in relation to behavioral research. The use of the linear program for cross-validational purposes is presented. Problems regarding replication with second-order functions are also delineated.

The use of co-variables utilizing the generalized analysis of variance model is presented in Chapter VII. The distinction between traditional co-variance and the procedure presented in this chapter is contrasted.

The relationship of the linear regression procedures to other multivariate analytic methods, such as multiple discriminant analysis and canonical analysis is also discussed in Chapter VII.

Chapter VIII presents specialized information which is not central to the presentation of the linear approach, but which might be helpful to the research investigator. For example, extended uses of sub-routine DATRAN are discussed. A few mathematical proofs regarding equivalences of (1) R^2 and variance of predicted scores, and (2) point biserial correlation, student's "t," and Fisher's "F" statistic, are presented.

The authors of this text wish to acknowledge indebtedness to Robert Bottenberg, Joe Ward jr, and Earl Jennings, who through their

N.D.E.A. Seminars have initiated our efforts and directed our early thinking regarding the regression approach. We also wish to thank Barbara Anderson, who labored over this manuscript.

FJK, DB, KMcN, JL, and TE, 1968

Contents

Chapter I Behavioral Science and Research Design

The models developed and used to explain human behavior are almost as varied as the numerous varieties of religious experience. Fortunately, at the base of behavioral models is an empiricism and an acceptance of observable data. This chapter summarizes a number of research findings and attempts to cast a generalized research model which can be used by all except possibly those who hold very extreme research positions.

Essentially, the only observable events we have are the response of the individual and some limited specification of the conditions under which he is responding. A portion of an individual's behavior can be explained by the conditions to which he is exposed. For example, (1) shine a bright light into the non-damaged eyes of an aroused individual and pupillary constriction is noted; or (2) introduce a sharp, loud noise and one can note a startle response and a digital vascular constriction.

Other behaviors, especially those which involve an extended delay (5 seconds or more) after stimulus presentation, cannot be easily attributed to the stimulus properties. For example, present an eight-year-old child with the visual stimuli: 2 + 4 + 6 = ?, and you may get a variety of verbal responses, i.e., 12, 8, etc. From such data we infer some mediational factor, such as an ability to add three digits. Other research leads us to postulate mediational factors which include: anxiety, motivation, expectancies, achievement (special learnings), and ability (convergent, divergent, spatial, verbal, etc.).

In one sense, the task of behavioral research is to determine functional relationships between human behavior (some quantified response) and the characteristics of the behaving individual. A few examples are given below.

If we know that one child expects his parents to hit him every time he enters the living room and another child expects to be hugged by his parents when he enters, we can predict a number of responses for each child when he enters the living room. One can say that, due to his reinforcement history, the child has a tendency to respond in such a manner in the presence of such stimuli. With knowledge of some expectancy, we can assign the number "1" to a child with expected hug and "0" to one with expected hit. We can count the number of smiles upon entering the room and note the correlation between category (expectancy) and the response. If we use knowledge of this sort, and if we know that strong relationships exist, we can predict outcome states with some confidence. Of course, this is a simple-minded example.

Sarason and Palola (1960) found that performance on coding tasks similar to coding tasks some intelligence tests use depends, in some degree, upon the subject's level of test anxiety as measured by the TAS. On difficult coding tasks, the high test-anxious Ss performed less adequately than did the low test-anxious. Conversely, given an easy coding task, the high test-anxious Ss are superior to low test-anxious Ss. These data reflect an interaction (multiplicative function) between task difficulty and test anxiety. Cast into a predictive equation we can get:

$$\tilde{Y}_c = a_1TA + a_2T + a_3(TA*T)$$

The predicted coding behavior ($\tilde{Y}_c$) of an individual is related to some weighted combination (a_1, a_2, a_3) of his level of test anxiety (TA), the difficulty of the coding task (T), and the interaction of these two variables (TA*T). Studies of this sort using only a few predictor variables usually yield predicted scores which introduce a fairly large amount of error (the difference between the predicted score and the actual observed score). These studies do, however, provide some notion regarding factors relevant to specific behaviors. Nevertheless, if we are to be successful in predicting and controlling significant human behavior, we apparently must introduce other variables which might be relevant to reducing error. A critical review of behavioral literature reveals that a network of interrelated variables are operating in any behaving context. These variables can be grouped into three categories: focal stimulus variables, person variables, and context variables.

Focal Stimulus Variables

The notion that human behavior is a function of both focal stimulus properties and individual characteristics has been with us a long time. The accumulated history of psychological research has specified a great number of sensory limens and stimulus response relationships. Early animal learning studies focused upon discrimination and generalization of stimulus properties. More recent psychological theorizing which centers upon focal stimulus properties deals with more complex aspects of the S-R relationship. Gagné (1965) has proposed a learning structure for number operations which hypothesizes a learning hierarchy that is composed of complex S-R chains. Guilford (1950, 1959, 1966) reported a number of content characteristics which might be relevant to problem solutions. Suppes (1966) has investigated process characteristics of arithmetic problems (e.g., the number of transformation, operation, and memory steps) which accounts for variation in success latency and proportion correct across problems.

Person Variables

Concurrent with studies specifying the effects of focal stimuli upon response, a body of data has been gathered regarding individual characteristics. Studies of the intellect from Binet to Guilford (1966) show that knowledge of individual "intellective functions" can *aid* prediction of performance across a broad class of tasks. Cattell (1966) reported a number of personality characteristics (e.g., ego strength, excitability, etc.) which aid prediction of school achievement over and above that which intelligence measures predicted. Manifest Anxiety (Taylor, 1953), Test Anxiety (Sarason, 1957), and n Achievement (McClelland, *et al*., 1953) have been reported to be related to complex behavior such as school success. Furthermore, refinements upon the anxiety studies show that anxiety level interacts with ability (Katahn, 1966; Denny, 1966) and task difficulty (Sarason and Palola, 1960; Castaneda, *et al.*, 1956).

Context Stimulus Variables

A number of non-task related stimuli can influence the student's behavior. Whiting and Child (1953) and McGuire, *et al.*, (1956) have reported that peer and adult expectancies are related to school success. In a sense, the context provides information regarding "pay-off" or reinforcement associated with performance.

In a comparison of an innovative school with control schools, Foster (1968) hypothesized that a number of biases in each program might theoretically interact with student characteristics to produce lawful variation in achievement. For example, an innovative school procedure which includes a large amount of independent study from the outset might result in exceptional academic growth for the bright autonomous student and only mediocre growth for the bright *low* autonomous student. Conversely, the bright *low* autonomous student might perform exceptionally well when the structure of the school is highly organized and supervised.

In the educational setting other context variables which might interact with student characteristics to influence response acquisition might be:

(1) Peer expectations: If the student's friends value school success, he might work even though he does not value school achievement.

(2) School "reward systems": Peer expectations can be manipulated by means of providing science "fairs," use of

academic games, etc., which "reward" students' academic behavior.

A Comprehensive Model for the Study of Complex Behavior

In view of the preceding discussion, a comprehensive model is presented in Figure 1-1. The figure is essentially an extension of the work of McGuire (1960) and Whiteside (1964).

The quasi-mathematical equation should read: The predicted behavior of individual "a" ($\tilde{Y}_a$) is a function of his personal characteristics (P), the focal stimuli characteristics (S_f), and characteristics of the context (S_c). The enumerated variables are only suggestive of relevant variables within each category.

The functional relationships among the three categories and the criterion (Y_a) can take many forms depending upon research expectations of the investigator. This text is designed to show you how complex functions can be statistically expressed using multiple linear regression analysis. It is premature to delineate these relationships in detail; however, the following example is provided for illustrative purposes.

Accounting for Complex Behavior

Consider the task of predicting complex behavior, such as success on a job-training program. In view of the proposed model, a number of aspects of the situation should be investigated. We might first examine the set of behaviors that are related to the criterion of success (Y_a). What are the focal stimuli characteristics (S_f)? Such an examination might give the investigator a

$$\tilde{Y}_a = f(P, S_f, S_c)$$

Where:

P = WITHIN PERSON VARIABLES

1. Convergent Thinking (California Test of Mental Maturity, Gestalt Transformation, and STEP Listening).
2. Divergent Thinking (Seeing Problems, Unusual Uses, and Consequences).
3. Symbol Aptitude (Mutilated Words, D.R.T. Reaction Time, Dotting).
4. Elements of Motivation (Surgency vs. Desurgency, nAch, Scholastic Motivation, etc., e.g., 16 PF, CPI, Test Anxiety, and the Academic Achievement Potential Questionnaire).
5. Sex-Role Identification (Masculinity-Femininity Scale, or designation of sex).
6. Special Learnings (relevant for success to the task at hand, e.g., specific responses lower in the hierarchy necessary for higher order response pairing).

S_f = CHARACTERISTICS OF FOCAL STIMULI (the task to be mastered)

1. Discriminative (stimulus characteristics).
2. Process Structures (e.g., number of transformation steps as delineated by Suppes, 1966, and sub-components which the structure encompasses as indicated by Gagné, 1965).

S_c = CHARACTERISTICS OF THE CONTEXT (non-task related stimuli which might influence the behavior under investigation)

1. School Structure (e.g., time in class, amount of independent study, special resources available).
2. Peer Expectancies (measures, i.e., Cellura's A.A.P.Q. (1968), in which students indicate the relative importance of academic and other success).
3. Adult Expectancies (social class factors and recognition for academic excellence, e.g., number of awards, academic games, science fairs, etc., all which might interact with and/or modify elements of motivation).

$\tilde{Y}_a$ = PREDICTED BEHAVIOR OF INDIVIDUAL "a"

Figure 1-1. One possible comprehensive model for study of complex behavior.

few notions regarding relevant human characteristics (P) needed for criterion success (e.g., are specific abilities and/or special previous learnings prerequisite?). Furthermore, what are the conditions

(S_c) surrounding the learning and testing setting? Is there a great deal of peer pressure for success? If so, will test anxiety be relevant? What are the intermediate pay-off schedules? Must the trainee work through the program before some reinforcement is given? Will the need to achieve be relevant to the reinforcement schedule? If he has failed often in the past, does he have an expectancy to fail in this new task? If so, can the training context be manipulated to minimize these effects? Familiarity with the univariate research literature, as well as with the multivariate taxonomies provided by Cattell (1966), Guilford (1966), etc., should provide additional suggestions regarding the relevant variables which account for complex behavior.

We might conclude that the desired terminal behavior of the training program is related to: (1) Spatial abilities (SA); (2) Need to achieve (nAch); (3) Anxiety level (Ax); (4) Ability to manipulate symbols (Sy); (5) Ability to read "effectively" (Ar); and (6) Past learning success (such as high school grade point average (GPA), etc.). This can be cast as:

$$\tilde{Y}_T = \int(SA, nAch, Ax, Sy, Ar, GPA)$$

where:

$\tilde{Y}_T$ = predicted terminal behavior used for evaluation.

We still have not specified the functions. We may assume the scores on the training task (terminal behavior) are a sum of the weighted six within-trainee variables (an additive function).

Equation 1.1

$$\tilde{Y}_T = a_1SA + a_2nAch + a_3Ax + a_4Sy + a_5Ar + a_6GPA$$

The weights $a_1 \ldots a_6$ might rationally be chosen or empirically derived using a least squares solution or some other solution. (Later, you will note that the multiple regression analysis we use will employ the least squares solution.)

Knowledge of past research might suggest to the investigator that the additive function is really not quite adequate. He might conclude that GPA is geometrically related to $\tilde{Y}_T$ (i.e., GPA's from the very low to the middle are not predictive; but those from the middle to the high are predictive). Furthermore, our hypothetical investigator might expect anxiety (Ax) and ability to manipulate symbols (Sy) to interact such that anxiety might adversely influence terminal behavior only when the trainee is low on symbol aptitude. These circumstances can be reflected by including a squared function of GPA and a multiplicative function (Ax*Sy). The expanded equation would be:

Equation 1.2

$$\tilde{Y}_T = a_1SA + a_2nAch + a_3Ax + a_4Sy + a_5Ar + a_6GPA + a_7(GPA)^2 + a_8(Ax*Sy)$$

(The asterisk (*) = multiplication. We use this convention since it is the FORTRAN symbol for this function.)

Theoretically, if the two new variables represented the functional relationship more adequately, then the predicted behavior ($\tilde{Y}_T$) should be very close to the actual behavior (Y_T), thus producing only a small amount of error in the prediction.

Equation 1.2 may seem rather large and bulky; yet it is an attempt to reflect the variables and functions which theoretically should influence the hypothetical terminal behavior under investigation.

Indeed, the intent of these few comments was to illustrate the problems which face the investigator when he encounters man in his natural environment. Apparently, if we are to reduce errors of prediction $(Y-\tilde{Y})$ or reduce error variance $\left(\frac{\Sigma(Y-\tilde{Y})^2}{N}\right)$, we need to construct comprehensive models of behavior. Furthermore, the multiple linear regression analysis presented in this book provides one statistical basis for testing these comprehensive models.

Chapter II Sampling Theory and Inferential Statistics

The research discussed in this book is based on the assumption that a researcher can make fairly accurate statements about a large group by examining a smaller group (representative of the larger group). A statistical _population_ is a collection of individuals or elements about which one is seeking information. Seldom, though, does the researcher have an opportunity to observe all individuals in a population; usually the researcher can observe only a relatively small fraction of the statistical population. This fraction or a subset of a statistical population is called a _sample_ of the population. Facts about the population are known as _parameters_, and facts about a sample are known as _statistics_. An important use of statistics is in the making of inferences about larger groups (populations) on the basis of information obtained from smaller groups (samples). The extent to which this inference can be made with any accuracy depends on the adequacy of our sample or samples.

Methods of Sampling

If one is to infer from a sample or samples to a population, he would like his sample to approximate the population closely with respect to the criteria being investigated. Quite often an experimenter suggests that the sample he investigated was drawn at random from the population. The word "random" can be used in one of three ways. (1) It can mean that the sample was drawn in a haphazard or unorganized manner. (2) The term may be used to describe certain methods or operations for getting a sample. For example, there are certain techniques used in games of chance (lotteries, sweepstakes, etc.) that hopefully prevent the biased selection of a sample. (3) The third possibly meaning of the word is that each subject or element in the population had an equal chance of being selected in the sample. The third definition, the more theoretical, is an assumption that underlies the derivation of many formulas used in sampling or inferential statistics. In most practical research settings, each individual subject or element in the population is not given an equal chance to be included in the sample, but practical operational methods are frequently used to approximate this assumption underlying the formulas.

If one is dealing with a finite population, such that all elements can be listed, a random sampling procedure can readily be applied to select a sample of the required size. The elements could be drawn from a box in such a manner that the names of the elements are not known to the individual doing the drawing. Another technique would be to assign code numbers to each element of the population and use a table of random numbers to select the elements to be included in the sample. Still another technique for selecting the sample would be to arrange the names in some order, for example the oldest to the youngest, and select every

k^{th} person to be a member of the sample. This technique has, on occasion, been misused and thus has led to wrong predictions (the 1948 presidential prediction), if all of the population can be listed, then the sample may be viewed as random.

A modification of random sampling is stratified random sampling. This procedure requires that the researcher have prior knowledge about the population with respect to various categories or strata. For instance, it might be important to stratify your sample with respect to males and females. If this is true, the researcher must know the proportion of males and females in the population and restrict the proportions in the sample according to those of the population. This technique has sometimes been called proportional stratification.

Although the technical definition of randomness is assumed in mathematical derivations of formulas, it must be recognized that insisting on rigorous random sampling methods would make many research works impossible.

Sampling Errors

The difference that can be found to exist between a parameter (population fact) and a statistic (sample fact) is called a sampling error. The concept of error implies that a parametric or true value can meaningfully be defined. Without an implication of a true value, the concept of error has no meaning. The reader should note that the definition of sampling error does not in any way indicate that a sample drawn at random is free of sampling error. It is important to realize that this error does exist in most samples; and the researcher must concern himself with the problem of estimating his error. This problem has been approached experimentally and will be discussed in the following section.

Sampling Distribution

The concept of sampling distributions is important to the researcher in terms of providing him with some evidence as to how well he can infer to the population from his sample. In practice, inferences are made from statistics that are calculated from a sample drawn at random from the population. If we continue to draw samples of the same size from the population, we could expect to find some of the facts about each sample to be different. These differences would give us some indication of the sampling error. This observation has led to the concept of sampling distributions.

> A sampling distribution is a theoretical probability distribution of the possible values of a sample fact which would occur if one were to draw all possible samples of the same size from a given population.

If the sampling distribution of the sample fact (statistic) is normal, the mean of the distribution is the expected value (similar to the parameter) of the statistic, and the standard deviation of the sampling distribution is the standard error of the statistic. In general, the smaller the standard error, the more confidence we have in our inference.

Steps in Hypothesis Testing

To this point we have been discussing the importance of random sampling, and we have discussed some of the methods for collecting or determining a sample. We can now incorporate sampling theory into the steps in hypothesis testing. These steps are:

STEP I: A statement of the research hypothesis. The research hypothesis is the actual question that the researcher is seeking to answer. This hypothesis is in no way inhibited by the nature of the statistical procedure.

STEP II: A statement of the statistical hypothesis. In many instances the hypothesis is the same as in STEP I. On occasion the research question must be redefined in statistical terms in order that the best techniques that are available can be used in assisting the researcher in answering the research hypothesis. For example, the researcher may hypothesize that one technique of teaching reading is more effective than another. In order to use the best statistical technique, the statistical hypothesis that the two techniques are equally effective is tested. In this instance the research hypothesis is rejected if we retain the statistical hypothesis.

STEP III: Determine the level of significance. The level of significance of a statistical test defines the probability level that the researcher considers too low to warrant support of the hypothesis being tested. If the probability of the obtained results is less than that determined before the data is examined, the researcher will not accept his <u>statistical</u> hypothesis.

The selection of the level of significance is not a straightforward procedure. The researcher must take into account the probability of both a Type I and a Type II error. A Type I error is the likelihood of rejecting a hypothesis that is true. The probability of making a Type I error is equal to the level of significance. A Type II error is the likelihood of retaining a hypothesis that is false. The researcher does not have complete control over this type of error, but the smaller the level of significance, the greater the chance of making a Type II error. (The concept of power is related to the Type II error, but a discussion of this concept is beyond the scope of this text.)

The researcher must decide which of the two types of errors is more serious, and then determine the level of significance.

STEP IV: Collect the data using appropriate sampling techniques and perform the designated statistical analysis (The t and F statistical tests will be briefly discussed at the end of this chapter.)

STEP V: Compare the results of the statistical test with the critical region of the statistic as determined by the level of significance in STEP III. If the result falls in the critical region, reject the statistical hypothesis and retain as tenable an alternative. If the result does not fall in the critical region, retain the statistical hypothesis as tenable. In some instances the researcher will have to go one step further and use the results of the statistical hypothesis to help answer the research hypothesis.

The t and F Statistics

Two statistics are frequently referred to throughout statistical texts. The t and F statistics are used quite often in inferential statistics. The two statistics will be discussed independently and then the relationship between the two statistics will be discussed.

The most common use of the t-test is to test the hypothesis that the means of two groups are equal. A standard formula for t is as follows:

$$t = \frac{(\bar{X}_1 - \bar{X}_2) - (\mu_1 - \mu_2)}{S_{\bar{x}_1 - \bar{x}_2}}$$

t = the value by which the statistical significance of the mean difference will be judged.

$(\bar{X}_1 - \bar{X}_2)$ = the observed difference between the means in the sample selected.

$(\mu_1 - \mu_2)$ = the hypothesized difference. (Usually hypothesized to be zero.)

$S_{\bar{x}_1 - \bar{x}_2}$ = the standard error of the difference between means.

There are four assumptions underlying the t-test:

1. The sample (samples) has (have) been randomly selected from the population.
2. The treatment populations are normally distributed with respect to the criterion.
3. The variability of the two treatment populations is the same (homogeneity of variance).
4. The means of the two populations are equal.

It is obvious the fourth assumption is the hypothesis being tested. If the first three assumptions are adequately met, the fourth assumption is the assumption which dictates the result. If the fourth assumption is true, the ratio is distributed as a t-distribution. If the fourth assumption is not met, the ratio is not distributed as the t-distribution, and this will provide the researcher with statistical significance.

The F-statistic is a technique for testing the difference among means of two or more groups. The F-statistic may be expressed in many ways, but we will discuss it from the following point of view:

$$F = \frac{\text{observed variance of treatment means} \;/\; df_1}{\text{expected chance variance of these means} \;/\; df_2}$$

(df_1 and df_2 refer to degrees of freedom and will be discussed later.)

The numerator of this ratio is determined by comparing the means of the groups with the overall mean or determining the variance among treatment means. The denominator is dependent on the variability that is found within the treatment groups. The value of the denominator increases as the variability of the groups increases. This indicates the more the groups vary within themselves, the more different one would expect the means to be.

Seldom is the ratio found to be less than one. The magnitude of F is the primary concern of the researcher. If the ratio is statistically larger than expected, this would mean that the observed variance among the means could not be attributed to the variance within the groups. If the obtained F-statistic is not significantly different from the F-distribution, this means that the variance found among the means can be accounted for by the variance within the groups.

The assumptions underlying the F-test are similar to those of the t-test (the number of groups changes). These assumptions are discussed at greater length at the end of Chapter IV. Again the first three assumptions must be met before the technique is acceptable.

As can be clearly seen, the t-statistic is a special case of the F-statistic. Mathematically, it has been shown that t^2 is equivalent to an F-statistic with two groups. It is for this reason the authors will use the F-statistic exclusively in the remainder of this text.

Chapter III Vectors and Vector Operations

Introduction

The multiple linear regression model, as well as almost any other statistical model, can be most easily understood within the framework of vector notation. If a researcher is to develop a command over the linear regression model, he must become relatively familiar with vector notation and vector operations. Once the researcher has grasped the basic components of vector algebra, he can construct for himself the statistical test used to test the particular research question. Thus, knowledge of vector algebra, coupled with mastery of the linear regression approach, allows the researcher to ask the research question in the particular way he wants and to then generate the statistical test. The present approach may seem heretical to those researchers who were taught that they must frame their research question in a particular fashion, or that only certain research questions were, in fact, permissible.

Most of us come into contact with vectors in our everyday life; and thus, the mastery of vector algebra is not really a difficult chore. Most of the properties of vector algebra are simply generalizations from the algebra with which we are already familiar.

Why Use Vectors?

Vectors allow us to represent data in a very concise fashion. Once the conventions are learned and the symbols are identified, a large amount of data can be represented in a small amount of space. An example of vectors with which we are all familiar is:

City	High Temperature (in degrees Fahrenheit)
Chicago	20
Miami	80
New York	70
St. Louis	30

Table 3-1. Example of a single vector; the same measurement for each of several entities

The information contained in Table 3-1 is familiar to all of us. The same kind of measurement has been made on each of the entities represented, i.e., the high temperature for a specified time-period has been recorded for each of the cities listed. The collection of temperatures is called a vector because each member of the collection is, in fact, a number; and the members (or "elements" or "components") are in some particular order. That is, the temperature of 20° was observed in Chicago, and the temperature of 30° was observed in St. Louis.

Height (in inches)	68
Weight (in pounds)	150
Age (in years)	25
Schooling (in years)	19
Married (yes = 1; no = 0)	1
Children (total number)	1

Table 3-2. Example of several measurements on a single entity.

The data in Table 3-2 indicates another use of vectors, representing several measurements on a single entity. The above information could have come from a biographical data sheet. If we are given the following vector (Y) and told that this vector represents the same information as presented in Table 3-2, then we would know that this individual is six feet tall, weighs 300 pounds, is thirty years old, has a high school education, is not married, and of course, has no children.

$$Y = \begin{bmatrix} 72 \\ 300 \\ 30 \\ 12 \\ 0 \\ 0 \end{bmatrix}$$

The data in Table 3-3 represents a combination of the kinds of vectors previously described. Here we have several observations on each of several entities. For the purposes of this book, we shall focus on vectors which represent the same kind of measurement on a number of subjects. Since there are four pieces of information about each entity or subject in Table 3-3, we will treat Table 3-3 as having four vectors. Someone has collected three pieces of information and then calculated a fourth piece of information (batting average) from two of the observations (dividing "hits" by "at bats").

Player	At Bats	Hits	Batting Average	Home Runs
Joe	120	60	.500	3
Sam	100	30	.300	5
Dick	110	22	.200	15
Jim	100	25	.250	20
Russ	120	48	.400	25

Table 3-3. Example of several vectors for each entity

Definition of Vector

The definition of a vector that we will use is simply: A vector is an ordered set of numbers. The number of elements in a vector is called the dimension of a vector. With reference to Table 3-3, and given the order of the players, we know that Joe hit 3 home runs because Joe is the subject corresponding to the first element of each of the four vectors; and we can readily see that the numerical value of the first element of the "home run" vector is 3. Likewise, since .400 is the fifth element of the "batting average" vector, and Russ is the subject corresponding to the fifth element, we can ascertain that Russ's batting average is .400.

We want to emphasize here that any number can be an element of a vector. We have already used positive numbers, zero, and decimal fractions. Negative numbers and common fractions are also valid candidates as elements of vectors.

The other property of a vector, that there is some order, should not be taken lightly. The order does not have to be inherent in the data; in fact, usually it is not. Referring again to Table 3-3 and the "hits" vector, there seems to be no order. The numbers are not arranged from high to low nor from low to high. The reader should be well aware, though, that the order has already been defined by the first column of names. There is numerical order in the vector representing home runs, but these were the actual observed home runs of the various players.

Vector Notation

The general notation that we will use is to represent vectors with subscripted capital letters and to represent elements of a vector by double subscripted capital letters. The particular subscripted letter which we associate with a vector is purely arbitrary but will sometimes have some mnemonic advantages. We could represent the biographical vector of Table 3-2 as B_1. B_{1_1} would then be the number corresponding to the subject's height (68), B_{1_2} would be the number corresponding to the subject's weight, and so on. The second subscript thus refers to the position of the element in the vector. In general terms, the vector W_1 with t elements can be represented by:

$$W_1 = \begin{bmatrix} W_{1_1} \\ W_{1_2} \\ W_{1_3} \\ \cdot \\ \cdot \\ \cdot \\ W_{1_t} \end{bmatrix}$$

where the three dots mean "and so on continuing to." These three dots are necessary because we do not know the value of t; there could be 50 elements of W_1, or 100, or just 4.

Example 3.1

If we represent the vector $\begin{bmatrix} 1 \\ 5 \\ 6 \\ 13 \\ -2 \end{bmatrix}$ by X_1,

then X_{1_1} is equal to 1, X_{1_2} is equal to 5, X_{1_3} is 6, X_{1_4} is 13, and X_{1_5} is -2. Note that X_{1_6}, X_{1_7}, etc., are not defined for this vector.

Categorical Vectors

We will find many occasions later on to use what some people call "categorical vectors." These vectors will usually assist us in representing group membership. We will assign one common number to entities which belong to the group and another common number to those entities which do not belong to the particular group under consideration. Any two numbers would do the job for us, but we find it most convenient to use a "1" if an entity is a member of the group under consideration and a "0" if the entity is not a member of the group under consideration. We cannot over-emphasize the fact that the judicious use of "1's" and "0's" for representation of group membership is one of the most crucial aspects of linear regression.

Example 3.2

Let us imagine that you are interested in studying the college student population and are interested in three variables: sex, college status, and marital status. With respect to the behavioral model introduced in Chapter I, you have reason to believe that the behavior you are investigating is a function of sex, college status, and marital status. In order to complete the example, let us assume that the behavior under investigation is the degree of one's liberal attitudes towards political ideas. Now let us construct the group membership vectors. The vectors that we need for this problem are:

G_1 = 1 if person is a male; and 0 otherwise.

G_2 = 1 if person is a female; and 0 otherwise.

G_3 = 1 if person is a freshman; and 0 otherwise.

G_4 = 1 if person is a sophomore; and 0 otherwise.

G_5 = 1 if person is a junior; and 0 otherwise.

G_6 = 1 if person is a senior; and 0 otherwise.

G_7 = 1 if person is married; and 0 otherwise.

G_8 = 1 if person is not married; and 0 otherwise.

Note that we used two vectors to represent sex and two vectors to represent marital status, while we used four vectors to represent college status. Another investigator might not be satisfied with our particular group designations. An additional marital status group could well be that of "divorced." This latter group could well increase the "usefulness" of the resulting function and thus increase the predictability of the behavior under consideration. It is up to the researcher to define his groups that he wants to include in his research. The linear regression approach is flexible in that it will handle as many groups as the researcher is willing to define and obtain data from. Even as innocuous a variable as sex can be divided into more than two groups if we are willing to re-define the variable as "sex-role interests" and divide the continuum of responses to such a questionnaire into, say, three or four groups.

Given the following vectors:

$$\begin{matrix} G_1 & G_2 & G_3 & G_4 & G_5 & G_6 & G_7 & G_8 \\ \begin{bmatrix}1\\0\\1\\1\end{bmatrix} & \begin{bmatrix}0\\1\\0\\0\end{bmatrix} & \begin{bmatrix}1\\0\\1\\0\end{bmatrix} & \begin{bmatrix}0\\0\\0\\0\end{bmatrix} & \begin{bmatrix}0\\0\\0\\0\end{bmatrix} & \begin{bmatrix}0\\1\\0\\1\end{bmatrix} & \begin{bmatrix}0\\1\\1\\1\end{bmatrix} & \begin{bmatrix}1\\0\\0\\0\end{bmatrix} \end{matrix}$$

We know at once that the first person (the first element in each vector) is a male, freshman, and unmarried. Inspection of the second element in each vector indicates that this is a female senior who is married.

Elementary Vector Algebra Operations

Equality of Two Vectors. Two vectors, F_2 and G_3, are said to be equal if $F_2 = G_3$ for all elements, or more precisely, $F_{2_i} = G_{3_i}$; $i = 1, 2, 3, \ldots t$; where t is the total number of elements of both F_2 and G_3. A very important implicit requirement is that both vectors have the same number of elements before equality can hold.

Example 3.3

$$A_4 = \begin{bmatrix}1\\3\\0\\4\\2\end{bmatrix} \quad B_5 = \begin{bmatrix}1\\3\\0\\4\\2\end{bmatrix} \quad C_3 = \begin{bmatrix}1\\3\\0\\4\\2\end{bmatrix} \quad D_5 = \begin{bmatrix}-1\\3\\0\\4\end{bmatrix} \quad A_6 = \begin{bmatrix}1\\3\\0\\2\\4\end{bmatrix}$$

$A_4 = B_5$ because $A_{4_1} = 1 = B_{5_1}$; $A_{4_2} = 3 = B_{5_2}$; $A_{4_3} = 0 = B_{5_3}$; $A_{4_4} = 4 = B_{5_4}$; and $A_{4_5} = 2 = B_{5_5}$.

$A_4 \neq C_3$ because $A_{4_1} \neq C_{3_1}$.

$A_4 \neq D_5$ because $A_{4_1} \neq D_{5_1}$; and the number of elements in the two vectors is not the same.

$B_5 \neq C_3$ because $B_{5_1} \neq C_{3_1}$.

$B_5 \neq D_5$ because $B_{5_1} \neq D_{5_1}$; and the number of elements in the two vectors is not the same.

$C_3 \neq D_5$ because the number of elements in the two vectors is not the same.

$A_4 \neq A_6$ because $A_{4_4} \neq A_{6_4}$; and $A_{4_5} \neq A_{6_5}$.

Addition of Two Vectors. The addition of two vectors is defined as the addition of each element in one vector to the corresponding element in the other vector. As with the equality of two vectors, an implicit requirement is that both vectors have the same number of elements before addition is possible. The addition of two vectors, A_2 and B_3, to produce the vector C_5 is written symbolically as:

$$C_5 = A_2 + B_3 = \begin{bmatrix}A_{2_1}\\A_{2_2}\\A_{2_3}\\\cdot\\\cdot\\\cdot\\A_{2_t}\end{bmatrix} + \begin{bmatrix}B_{3_1}\\B_{3_2}\\B_{3_3}\\\cdot\\\cdot\\\cdot\\B_{3_t}\end{bmatrix} = \begin{bmatrix}A_{2_1}+B_{3_1}\\A_{2_2}+B_{3_2}\\A_{2_3}+B_{3_3}\\\cdot\\\cdot\\\cdot\\A_{2_t}+B_{3_t}\end{bmatrix} = \begin{bmatrix}C_{5_1}\\C_{5_2}\\C_{5_3}\\\cdot\\\cdot\\\cdot\\C_{5_t}\end{bmatrix} = C_5$$

Example 3.4

$$A_4 = \begin{bmatrix} 1 \\ 0 \\ 3 \\ -4 \end{bmatrix} \qquad B_4 = \begin{bmatrix} 2 \\ 1 \\ 5 \\ 7 \end{bmatrix} \qquad D_4 = \begin{bmatrix} 3 \\ 0 \\ 4 \end{bmatrix} \qquad C_4 = A_4 + B_4$$

$$C_4 = A_4 + B_4 = \begin{bmatrix} 1 \\ 0 \\ 3 \\ -4 \end{bmatrix} + \begin{bmatrix} 2 \\ 1 \\ 5 \\ 7 \end{bmatrix} = \begin{bmatrix} 1 + 2 \\ 0 + 1 \\ 3 + 5 \\ -4 + 7 \end{bmatrix} = \begin{bmatrix} 3 \\ 1 \\ 8 \\ 3 \end{bmatrix} = C_4$$

The addition of B_4 and D_4 is not possible because these two vectors do not have the same number of elements.

Subtraction of Two Vectors. The subtraction of two vectors is as straightforward as the addition of two vectors: If we want to subtract B_6 from A_2 to produce the vector C_5, we simply subtract corresponding elements of B_6 from A_2. $C_{5_i} = A_{2_i} - B_{6_i}$.

Example 3.5

$$A_2 = \begin{bmatrix} 2 \\ 0 \\ 7 \\ -4 \end{bmatrix} \qquad B_6 = \begin{bmatrix} 1 \\ 1 \\ 5 \\ 7 \end{bmatrix} \qquad C_5 = A_2 - B_6$$

$$C_5 = A_2 - B_6 = \begin{bmatrix} 2 \\ 0 \\ 7 \\ -4 \end{bmatrix} - \begin{bmatrix} 1 \\ 1 \\ 5 \\ 7 \end{bmatrix} = \begin{bmatrix} 2 - 1 \\ 0 - 1 \\ 7 - 5 \\ -4 - 7 \end{bmatrix} = \begin{bmatrix} 1 \\ -1 \\ 2 \\ -11 \end{bmatrix} = C_5$$

Multiplication of a Vector by a Number. We will find many occasions later on to multiply a vector by a number; and so we need to become familiar with this operation. The multiplication of a vector by a number is defined as the multiplication of each element of the vector by that number.

$C_6 = k*A_5$ (where k is a constant) is computed by multiplying each element of A_5 by the constant k.

$$C_6 = k*A_5 = \begin{bmatrix} k*A_{5_1} \\ k*A_{5_2} \\ k*A_{5_3} \\ \cdot \\ \cdot \\ \cdot \\ k*A_{5_t} \end{bmatrix} = C_6$$

Example 3.6

Let us suppose that we have made some weight observations on a number of entities, and it is desired to change the data from the original units of pounds to the new unit of ounces. Thus, we have to multiply each observation by the constant 16, because there are 16 ounces per pound. This operation can be represented in vector notation as:

$$Z_3 = 16*P_3$$

where the vector P_3 represents the original observations in terms of pounds, and the vector Z_3 represents the observations in the form of the new units. Suppose that our original vector looked like this:

$$P_3 = \begin{bmatrix} 2 \\ 4 \\ 6 \\ 3 \\ 0 \end{bmatrix}$$

Then the new vector Z_3 would be:

$$Z_3 = 16*P_3 = 16*\begin{bmatrix} 2 \\ 4 \\ 6 \\ 3 \\ 0 \end{bmatrix} = \begin{bmatrix} 16*2 \\ 16*4 \\ 16*6 \\ 16*3 \\ 16*0 \end{bmatrix} = \begin{bmatrix} 32 \\ 64 \\ 96 \\ 48 \\ 0 \end{bmatrix} = Z_3$$

The reader should note that it makes sense to multiply each and every element of the vector by the constant. The fourth element of vector Z_3, for instance, should represent (in ounces) the number of pounds corresponding to the fourth element of P_3. We note that the fourth element of Z_3 (48 ounces) is, in fact, equivalent to the fourth element of P_3 (3 pounds).

Example 3.7. We would like to introduce another example here to illustrate the simplicity and conciseness of vector algebra. Suppose that it is desired to change a 100-element vector of observations reported in units of feet to units in terms of inches. We can represent this operation with vectors in the following fashion: Where F_2 is the vector of observations in feet, I_2 is the new vector of observation in inches, and k is our multiplication constant (12 in this case because there are 12 inches per foot):

$$I_2 = k*F_2 = 12*F_2$$

Again, there is not just one multiplication inferred by the above expression; but every element of F_2 is multiplied by the constant k to produce the corresponding element in I_2.

Useful Properties of Vectors

Combining our recently learned knowledge of vectors with some previous knowledge about the properties of ordinary numbers, we can express the following useful properties:

Property 1: $X_3 + (-1)X_3 = X_3 - X_3 = \underline{0}$

where $\underline{0}$ represents a vector whose every element is equal to 0. Such a vector is often called the null vector.

$$\begin{bmatrix} X_{3_1} \\ X_{3_2} \\ X_{3_3} \\ \cdot \\ \cdot \\ \cdot \\ X_{3_t} \end{bmatrix} + (-1) \begin{bmatrix} X_{3_1} \\ X_{3_2} \\ X_{3_3} \\ \cdot \\ \cdot \\ \cdot \\ X_{3_t} \end{bmatrix} = \begin{bmatrix} X_{3_1} \\ X_{3_2} \\ X_{3_3} \\ \cdot \\ \cdot \\ \cdot \\ X_{3_t} \end{bmatrix} + \begin{bmatrix} -X_{3_1} \\ -X_{3_2} \\ -X_{3_3} \\ \cdot \\ \cdot \\ \cdot \\ -X_{3_t} \end{bmatrix} = \begin{bmatrix} X_{3_1}-X_{3_1} \\ X_{3_2}-X_{3_2} \\ X_{3_3}-X_{3_3} \\ \cdot \\ \cdot \\ \cdot \\ X_{3_t}-X_{3_t} \end{bmatrix} = \begin{bmatrix} 0 \\ 0 \\ 0 \\ \cdot \\ \cdot \\ \cdot \\ 0 \end{bmatrix} = \underline{0}$$

The above property becomes useful when we have occasion to subtract a vector from itself. The result of such an operation is simply a vector whose every element is equal to zero, or simply the null vector $\underline{0}$.

Property 2: $X_3 + \underline{0} = X_3$

$$\begin{bmatrix} X_{3_1} \\ X_{3_2} \\ X_{3_3} \\ \cdot \\ \cdot \\ \cdot \\ X_{3_t} \end{bmatrix} + \begin{bmatrix} 0 \\ 0 \\ 0 \\ \cdot \\ \cdot \\ \cdot \\ 0 \end{bmatrix} = \begin{bmatrix} X_{3_1}+0 \\ X_{3_2}+0 \\ X_{3_3}+0 \\ \cdot \\ \cdot \\ \cdot \\ X_{3_t}+0 \end{bmatrix} = \begin{bmatrix} X_{3_1} \\ X_{3_2} \\ X_{3_3} \\ \cdot \\ \cdot \\ \cdot \\ X_{3_t} \end{bmatrix} = X_3$$

This property tells us that if we desire to add the null vector ($\underline{0}$) to another vector, then the resulting sum will simply be the original vector. Thus, we do not change any element of a vector by adding the null vector to it.

Property 3: $a*(X_3+Y_4) = (a*X_3) + (a*Y_4)$

$$a(X_3+Y_4) = a\begin{bmatrix} X_{3_1} \\ X_{3_2} \\ X_{3_3} \\ \cdot \\ \cdot \\ \cdot \\ X_{3_t} \end{bmatrix} + \begin{bmatrix} Y_{4_1} \\ Y_{4_2} \\ Y_{4_3} \\ \cdot \\ \cdot \\ \cdot \\ Y_{4_t} \end{bmatrix} = a\begin{bmatrix} (X_{3_1}+Y_{4_1}) \\ (X_{3_2}+Y_{4_2}) \\ (X_{3_3}+Y_{4_3}) \\ \cdot \\ \cdot \\ \cdot \\ (X_{3_t}+Y_{4_t}) \end{bmatrix} = \begin{bmatrix} a(X_{3_1}+Y_{4_1}) \\ a(X_{3_2}+Y_{4_2}) \\ a(X_{3_3}+Y_{4_3}) \\ \cdot \\ \cdot \\ \cdot \\ a(X_{3_t}+Y_{4_t}) \end{bmatrix}$$

$$= \begin{bmatrix} aX_{3_1}+aY_{4_1} \\ aX_{3_2}+aY_{4_2} \\ aX_{3_3}+aY_{4_3} \\ \cdot \\ \cdot \\ \cdot \\ aX_{3_t}+aY_{4_t} \end{bmatrix} = \begin{bmatrix} aX_{3_1} \\ aX_{3_2} \\ aX_{3_3} \\ \cdot \\ \cdot \\ \cdot \\ aX_{3_t} \end{bmatrix} + \begin{bmatrix} aY_{4_1} \\ aY_{4_2} \\ aY_{4_3} \\ \cdot \\ \cdot \\ \cdot \\ aY_{4_t} \end{bmatrix} = a\begin{bmatrix} X_{3_1} \\ X_{3_2} \\ X_{3_3} \\ \cdot \\ \cdot \\ \cdot \\ X_{3_t} \end{bmatrix} + a\begin{bmatrix} Y_{4_1} \\ Y_{4_2} \\ Y_{4_3} \\ \cdot \\ \cdot \\ \cdot \\ Y_{4_t} \end{bmatrix} = aX_3 + aY_4$$

Property 3 tells us that if we desire to multiply the sum of two vectors by a constant (a), we can simply multiply each vector by the constant and then add the two resulting products.

Example 3.8. Suppose that we want to multiply the total GRE of five subjects by a constant of 3. The above section indicates that we can multiply both the verbal and quantitative sections by 3 and add these products; or we can first sum the verbal and quantitative sections and then multiply this sum by the constant 3.

$$V_3 = \begin{bmatrix} 200 \\ 250 \\ 300 \\ 400 \\ 500 \end{bmatrix} \qquad Q_3 = \begin{bmatrix} 400 \\ 450 \\ 500 \\ 600 \\ 700 \end{bmatrix} \qquad T_3 = V_3+Q_3 = \begin{bmatrix} 600 \\ 700 \\ 800 \\ 1000 \\ 1200 \end{bmatrix}$$

$$3*(V_3+Q_3) = (3*V_3) + (3*Q_3)$$

$$3*\begin{bmatrix} 600 \\ 700 \\ 800 \\ 1000 \\ 1200 \end{bmatrix} = \begin{bmatrix} 1800 \\ 2100 \\ 2400 \\ 3000 \\ 3600 \end{bmatrix} = 3*\begin{bmatrix} 200 \\ 250 \\ 300 \\ 400 \\ 500 \end{bmatrix} + 3*\begin{bmatrix} 400 \\ 450 \\ 500 \\ 600 \\ 700 \end{bmatrix} = \begin{bmatrix} 600 \\ 750 \\ 900 \\ 1200 \\ 1500 \end{bmatrix} + \begin{bmatrix} 1200 \\ 1350 \\ 1500 \\ 1800 \\ 2100 \end{bmatrix} = \begin{bmatrix} 1800 \\ 2100 \\ 2400 \\ 3000 \\ 3600 \end{bmatrix}$$

Property 4: $(a+b)*X_3 = (a*X_3)+(b*X_3)$

$$(a+b)*X_3 = \begin{bmatrix} (a+b)X_{3_1} \\ (a+b)X_{3_2} \\ (a+b)X_{3_3} \\ \cdot \\ \cdot \\ \cdot \\ (a+b)X_{3_t} \end{bmatrix} = \begin{bmatrix} aX_{3_1} + bX_{3_1} \\ aX_{3_2} + bX_{3_2} \\ aX_{3_3} + bX_{3_3} \\ \cdot \\ \cdot \\ \cdot \\ aX_{3_t} + bX_{3_t} \end{bmatrix} = \begin{bmatrix} aX_{3_1} \\ aX_{3_2} \\ aX_{3_3} \\ \cdot \\ \cdot \\ \cdot \\ aX_{3_t} \end{bmatrix} + \begin{bmatrix} bX_{3_1} \\ bX_{3_2} \\ bX_{3_3} \\ \cdot \\ \cdot \\ \cdot \\ bX_{3_t} \end{bmatrix} = aX_3 + bX_3$$

The above property tells us that when we want to multiply a vector by the sum of two constants, we can either add the two constants together and then multiply the vector by the sum of the two constants; or we can multiply the vector by the two separate constants and then add the resultant products. The former procedure is, in practice, much easier and quicker because fewer operations are involved.

<u>Property 5</u>: $(0)*X_4 = \underline{0}$

$$0*X_4 = 0*\begin{bmatrix} X_{4_1} \\ X_{4_2} \\ X_{4_3} \\ \cdot \\ \cdot \\ \cdot \\ X_{4_t} \end{bmatrix} = \begin{bmatrix} 0*X_{4_1} \\ 0*X_{4_2} \\ 0*X_{4_3} \\ \cdot \\ \cdot \\ \cdot \\ 0*X_{4_t} \end{bmatrix} = \begin{bmatrix} 0 \\ 0 \\ 0 \\ \cdot \\ \cdot \\ \cdot \\ 0 \end{bmatrix} = \underline{0}$$

Thus, multiplication of a vector by the number "0" yields the null vector, $\underline{0}$. No matter what the elements of a vector, if we multiply the vector by 0, we will end up with the null vector as our product.

<u>Property 6</u>: $(1)*X_3 = X_3$

$$(1)*X_3 = (1)*\begin{bmatrix} X_{3_1} \\ X_{3_2} \\ X_{3_3} \\ \cdot \\ \cdot \\ \cdot \\ X_{3_t} \end{bmatrix} = \begin{bmatrix} (1)*X_{3_1} \\ (1)*X_{3_2} \\ (1)*X_{3_3} \\ \cdot \\ \cdot \\ \cdot \\ (1)*X_{3_t} \end{bmatrix} = \begin{bmatrix} X_{3_1} \\ X_{3_2} \\ X_{3_3} \\ \cdot \\ \cdot \\ \cdot \\ X_{3_t} \end{bmatrix} = X_3$$

Thus, multiplication of any vector by the constant "1" yields the same vector.

These last two properties may seem trivial, and indeed they are quite trivial -- although very powerful and useful. Thus, the student is encouraged to understand and master these properties before proceeding. These properties of vector algebra, coupled with the concept to be discussed in the next section, form the structure of multiple linear regression. The more adept the student becomes with the concepts presented in the present chapter, the more adequately he can handle the building and simplification of linear regression models.

Linear Combinations of Vectors

Linear Combinations of Two Vectors. We will find many situations later on where we will be combining vectors and also where we will be determining if certain vectors are, in fact, linear combinations of other vectors. We, therefore, need to define the important concept of linear combinations.

Definition of a linear combination is as follows: Vector X_3 is said to be a linear combination of vectors Y_3 and Z_3 if there exists two numbers (constants), a and b (of which at least one is not zero), such that the following relationship holds:

$$X_3 = (a*Y_3) + (b*Z_3)$$

Example 3.9

Given: $X_3 = \begin{bmatrix} 3 \\ 4 \\ 5 \end{bmatrix} \qquad Y_3 = \begin{bmatrix} 1 \\ 2 \\ 3 \end{bmatrix} \qquad Z_3 = \begin{bmatrix} 1 \\ 0 \\ -1 \end{bmatrix}$

X_3 is a linear combination of Y_3 and Z_3 because:

$$X_3 = (a*Y_3) + (b*Z_3) \quad \text{when } a = 2 \text{ and } b = 1.$$

$$(a*Y_3) + (b*Z_3) = 2*\begin{bmatrix} 1 \\ 2 \\ 3 \end{bmatrix} + 1*\begin{bmatrix} 1 \\ 0 \\ -1 \end{bmatrix} = \begin{bmatrix} 2 \\ 4 \\ 6 \end{bmatrix} + \begin{bmatrix} 1 \\ 0 \\ -1 \end{bmatrix} = \begin{bmatrix} 3 \\ 4 \\ 5 \end{bmatrix} = X_3$$

<u>Some Special Linear Combinations of Vectors</u>. A total test score vector (computed by simply adding together the two subtest scores) is a linear combination of the two subtest vectors. In this instance, the weighting coefficients, a and b, are both equal to 1.

<u>Example 3.10</u>. Let us consider the example of total GRE (Graduate Record Exam), computed by simply adding together the verbal and quantitative subtest scores. Given the verbal vector (V_4) and the quantitative vector (Q_4):

$$V_4 = \begin{bmatrix} 400 \\ 500 \\ 600 \\ 350 \end{bmatrix} \qquad Q_4 = \begin{bmatrix} 400 \\ 300 \\ 400 \\ 750 \end{bmatrix}$$

The total GRE vector (T_4) is:

$$T_4 = [(1)*V_4] + [(1)*Q_4] = V_4 + Q_4$$

Therefore:

$$T_4 = \begin{bmatrix} 400 \\ 500 \\ 600 \\ 350 \end{bmatrix} + \begin{bmatrix} 400 \\ 300 \\ 400 \\ 750 \end{bmatrix} = \begin{bmatrix} 800 \\ 800 \\ 1000 \\ 1100 \end{bmatrix}$$

Another special case of a linear combination occurs when we multiply a vector by a number. The weight of the "second vector" is in this instance zero; and because it is zero, the elements of the second vector are of no consequence to us. That is, it doesn't matter what the elements of the second vector are because we already know that multiplication of any vector by zero will yield the null vector.

<u>Example 3.11</u>

$$A_3 = \begin{bmatrix} 4 \\ 3 \\ 0 \\ 1 \end{bmatrix} \qquad B_5 = \begin{bmatrix} B_{51} \\ B_{52} \\ B_{53} \\ B_{54} \end{bmatrix} \qquad C_6 = \begin{bmatrix} 24 \\ 18 \\ 0 \\ 6 \end{bmatrix} \qquad C_6 = (6*A_3) + (0*B_5)$$

$$C_6 = 6*\begin{bmatrix}4\\3\\0\\1\end{bmatrix} + 0*\begin{bmatrix}B_{5_1}\\B_{5_2}\\B_{5_3}\\B_{5_4}\end{bmatrix} = \begin{bmatrix}6*4\\6*3\\6*0\\6*1\end{bmatrix} + \begin{bmatrix}0*B_{5_1}\\0*B_{5_2}\\0*B_{5_3}\\0*B_{5_4}\end{bmatrix} = \begin{bmatrix}24\\18\\0\\6\end{bmatrix} + \begin{bmatrix}0\\0\\0\\0\end{bmatrix} = \begin{bmatrix}24\\18\\0\\6\end{bmatrix}$$

The vector C_6 is thus a linear combination of the vector A_3.

The concept of a linear combination of vectors is not restricted to just two vectors. A vector may be a linear combination of more than two vectors. The following example will help to clarify this point.

Example 3.12

$$A_4 = \begin{bmatrix}4\\3\\2\\1\end{bmatrix} \quad B_6 = \begin{bmatrix}1\\0\\0\\0\end{bmatrix} \quad C_9 = \begin{bmatrix}0\\1\\0\\0\end{bmatrix} \quad D_8 = \begin{bmatrix}0\\0\\1\\0\end{bmatrix} \quad E_7 = \begin{bmatrix}0\\0\\0\\1\end{bmatrix}$$

A_4 is a linear combination of B_6, C_9, D_8, and E_7 because:

$$A_4 = (4*B_6) + (3*C_9) + (2*D_8) + (1*E_7)$$

The reader should verify the above statement by actually carrying out the implied multiplication.

The vector B_6 is not a linear combination of vectors C_9, D_8, and E_7 because there does not exist any weighting coefficients such that

$$B_6 = (a*C_9) + b*D_8) + (c*E_7).$$

Mutually Exclusive Group Membership Vectors

Another special linear combination of vectors occurs when we add together mutually exclusive group membership vectors. Let us give an example.

Example 3.13. Suppose that we have occasion to deal with the variables of sex and marital status. Let us define our group membership vectors as follows:

S_1 = 1 if subject is a female; 0 otherwise.

S_2 = 1 if subject is a male; 0 otherwise.

M_1 = 1 if subject is married; 0 otherwise.

M_2 = 1 if subject is not married; 0 otherwise.

	S_1	S_2	M_1	M_2
Joe	0	1	1	0
Sam	0	1	1	0
Sue	1	0	1	0
Sally	1	0	0	1
Jane	1	0	0	1
Jack	0	1	0	1

S_1 and S_2 are mutually exclusive group membership vectors; that is, all subjects belong to one or the other categories of "male" and "female." Likewise, the categories of "married" and "not married" exhaust all of the possibilities of marital status (as far as the present researcher is concerned). We could have included other categories of marital status, but evidently the research question did not demand any additional categories.

One way of checking to see if the stated categories are, in fact, mutually exclusive is to compute the linear combination (all weights equal to 1) of the vectors under consideration. If these vectors are in fact mutually exclusive, then they will take into account each and every subject once and only once. That is, if we represent group membership vectors by "1's" and "0's," the resultant sum of the mutually exclusive group membership vectors will yield a vector with all elements equal to one (1).

A vector with all of its elements equal to 1 is called the unit vector and is symbolized as "U."

Let us add together the two sex vectors. Here we are computing a linear combination because the weights can be thought of as being equal to one (1) as in Example 3.10.

$$[(1)*S_1] + [(1)*S_2] = S_1 + S_2 = \begin{bmatrix}0\\0\\1\\1\\1\\0\end{bmatrix} + \begin{bmatrix}1\\1\\0\\0\\0\\1\end{bmatrix} = \begin{bmatrix}0+1\\0+1\\1+0\\1+0\\1+0\\0+1\end{bmatrix} = \begin{bmatrix}1\\1\\1\\1\\1\\1\end{bmatrix} = U$$

Thus, S_1 and S_2 are mutually exclusive because their sum is equal to the unit vector.

Let us now show that M_1 and M_2 are mutually exclusive vectors.

$$[(1)*M_1] + [(1)*M_2] = M_1 + M_2 = \begin{bmatrix}1\\1\\1\\0\\0\\0\end{bmatrix} + \begin{bmatrix}0\\0\\0\\1\\1\\1\end{bmatrix} = \begin{bmatrix}1+0\\1+0\\1+0\\0+1\\0+1\\0+1\end{bmatrix} = \begin{bmatrix}1\\1\\1\\1\\1\\1\end{bmatrix} = U$$

The reader should now have a good feeling for the fact that the unit vector can, in some cases, be considered as a linear combination of mutually exclusive group membership vectors. In fact, we have relied upon the unit vector to define mutually exclusive group membership vectors; so the above statement is simply a consequence of our definition -- but a very important consequence as we shall discover in later chapters.

The Concept of Linear Dependency

We will discover later that the concept of linear dependency is very important. The concept can easily be introduced at this point because it deals with linear combinations of vectors. A linear dependency occurs when one vector in a set of vectors can be expressed as a linear combination of the other vectors. Such a vector is said to be linearly dependent upon the other vectors. A linearly dependent vector, because it can be expressed in terms of other vectors, is redundant information and, as such, is actually not useful in terms of predicting behavior. There will be times, though, when we will want to include linearly dependent vectors, but we need to know when we have done so.

Example 3.14. Let us take another look at Example 3.8, which involved the total GRE and the two subtests. In that example we showed how the total GRE could be expressed as the sum of the two subtests. The reader should verify for himself that the verbal subtest (V_3) can be expressed as the total GRE (T_3) minus the quantitative subtest (Q_3); or $V_3 = (1)*T_3 + (-1)*Q_3$. It is also true that the quantitative subtest is a linear combination of the total GRE and the verbal subtest; or $Q_3 = (1)*T_3 + (-1)*V_3$.

Any one of the three vectors in Example 3.14 is linearly dependent on the other two because it can be expressed as a linear combination of the other two. The total GRE cannot be expressed in terms of the verbal subtest alone nor the quantitative subtest alone. That is, no weight (a) can be found such that: $T_3 = (a)*V_3$ nor such that $T_3 = (a)*Q_3$. Likewise, no weight (a) can be found such that: $V_3 = (a)*Q_3$. The reader should investigate the other possible combinations of vectors.

In Example 3.14 we end up with a set of two vectors of information which are said to be linearly independent.

> Definition of linear independence: A vector is said to be linearly independent if that vector cannot be expressed as a linear combination of the other vectors in the set.

If a vector can be expressed as a linear combination of the other vectors in the set, then it is redundant information and as such must be eliminated from the set of vectors when attempting to determine the number of linearly independent vectors in a set of vectors. The following example is intended to clarify the determination of the number of linearly independent vectors.

<u>Example 3.15</u>

$$V_1 = \begin{bmatrix} 1 \\ 2 \\ 3 \\ 4 \\ 5 \end{bmatrix} \qquad V_2 = \begin{bmatrix} 2 \\ 4 \\ 6 \\ 8 \\ 10 \end{bmatrix} \qquad V_3 = \begin{bmatrix} 4 \\ 8 \\ 12 \\ 16 \\ 20 \end{bmatrix} \qquad V_4 = \begin{bmatrix} 1 \\ 2 \\ 3 \\ 4 \\ 5 \end{bmatrix}$$

V_2 is a linear combination of the other vectors.

$$V_2 = (2)*V_1 + (0)*V_3 + (0)*V_4$$

or

$$V_2 = (2)*V_1$$

Therefore, V_2 is essentially eliminated from the set of vectors for the purpose of determining the number of linearly independent vectors in the set.

V_3 is a linear combination of the remaining vectors:

$$V_3 = (4)*V_1 + (0)*V_4$$

or

$$V_3 = (4)*V_1$$

Therefore, V_3 is essentially eliminated from the set of vectors for the purpose of determining the number of linearly independent vectors in the set.

We now have to consider the two remaining vectors, V_1 and V_4. The problem is to find a weight (a) such that: $V_1 = (a)*V_4$.

We note that the first four elements of V_4 must be multiplied by a weight of 1, whereas the fifth element must be multiplied by a weight of 1.25. Thus, there is no one weight which will suffice. Vectors V_1 and V_4 are thus linearly independent. We have, therefore, determined that there are two linearly independent vectors in the set of vectors in Example 3.15.

Problems for Chapter III

Let: $L_1 = \begin{bmatrix} 6 \\ 1 \\ 0 \end{bmatrix}$ $M_2 = \begin{bmatrix} 1 \\ 1 \\ 4 \end{bmatrix}$ $N_3 = \begin{bmatrix} -1 \\ 3 \\ -2 \end{bmatrix}$ $P_6 = \begin{bmatrix} 2 \\ 1 \\ 2 \end{bmatrix}$ $Q_5 = \begin{bmatrix} 1 \\ 0 \\ -2 \end{bmatrix}$

Compute the following:

(a) $M_2 + N_3$

(b) $N_3 + M_2$

(c) $2P_6 + Q_5$

(d) $Q_5 + 2P_6$

(e) $M_2 - L_1$

(f) $L_1 - M_2$

(g) Is M_2 a linear combination of P_6 and Q_5? Show your calculations in vector notation.

(h) How many linearly independent vectors are there in the following set? M_2, P_6, Q_5.

Chapter IV Analysis of Research Questions: Mutually Exclusive Categories

Introduction

So far, a comprehensive behavioral model has been presented (Chapter I), a few notions regarding sampling theory and inferential statistics have been introduced (Chapter II), and a few uses of vector notation were discussed (Chapter III). This chapter introduces the analysis of variance using multiple linear regression models. Since the intent of the chapter is to indicate the simple elegance of the model, the models introduced will not approach the complexities discussed in Chapter I. Of course, subsequent chapters will return to the complex models, which really are not so complex once the fundamental notions are mastered. Since the traditional analysis of variance techniques are a special case of regression analysis, we first present an example using traditional procedures; then we present the same example using a regression approach showing the identity between the two methods.

Traditional Analysis of Variance: Example I

In Chapter II, the first recommended step in hypothesis testing was to state a research hypothesis. Let us take a hypothetical case in which we investigate the influence of three different programed texts upon performance by a group of eleventh-grade boys on a physics examination.

We might hypothesize that the influence of treatment is equal; that is, there is no difference among treatment means.

The population of eleventh-grade boys, of course, is large; and to conduct a study with all eleventh-grade boys is impossible. Therefore, we might wish to use some sampling procedures and some inferential statistical procedure such as the analysis of variance.

We can visualize the total population to be something like Figure 4-1.

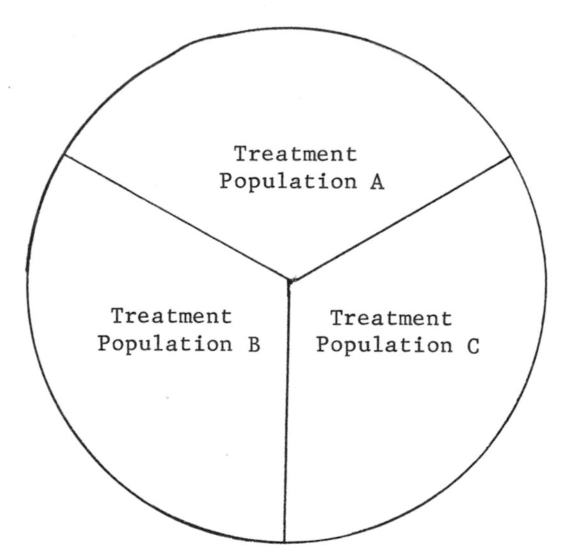

Figure 4-1. The three treatment populations represent subsets of the total population.

Suppose our sampling procedure is to give each person within each treatment population a number and then select subjects for treatment using a table of random numbers. (For a description of this procedure, see Edwards, 1964, p. 250.) This procedure satisfies the assumption of random selection.

> We shall assume that the three treatment populations are normally distributed with respect to the criterion and that the variance of the three treatment populations with respect to the criterion is the same. (The F-statistic has been found to be very robust, and the failure to meet these two assumptions has been found to have little effect on the probability of rejecting a true hypothesis, Type I error. See Box, 1954, or Norton, 1952.)

For purposes of exposition, we shall randomly assign two subjects per treatment (awfully small, but we just want to illustrate the parcelling of variances).

Upon completion of working through the different programed texts (treatments), we get the following physics test scores for members of each treatment:

Treatment A:	Subject 1 = 4
	Subject 2 = 6
Treatment B:	Subject 3 = 10
	Subject 4 = 12
Treatment C:	Subject 5 = 20
	Subject 6 = 20

The mean for Treatment A is [(4+6) / 2] = 5

The mean for Treatment B is [(10+12) / 2] = 11

The mean for Treatment C is [(20+20) / 2] = 20

The total mean is [(4+6 + 10+12 + 20+20) / 6] = 12

Now we want to know if the difference among group members is significant. Obviously 5 is not equal to 11, nor is 20 equal to 11 or 5, but remember that we have used samples and these mean differences might be due to sampling error. If you recall from

Chapter II, the F-statistic, through analyzing within-group variances and among-group variances, can suggest how frequently the observed mean differences are likely to occur.

$$F = \frac{\text{observed variance of treatment means} \;/\; df_1}{\text{expected chance variance of these means} \;/\; df_2}$$

Each value in the equation is discussed in the following order: (1) observed variance of treatment means; (2) expected chance variance of these means; (3) degrees of freedom (df); and (4) the F ratio.

Observed Variance of Treatment Means. To calculate the observed variance of treatment means, we first subtract each treatment mean from the total mean, and weight this value by the number of observations which comprise the group mean (in this example each group mean was based upon two individuals; therefore, the weight value is 2). Figure 4-2 summarizes this operation.

Treatment

$$A \quad \begin{bmatrix} 5-12 \\ 5-12 \end{bmatrix} \begin{matrix} = \\ = \end{matrix} \begin{bmatrix} -7 \\ -7 \end{bmatrix}$$

$$B \quad \begin{bmatrix} 11-12 \\ 11-12 \end{bmatrix} \begin{matrix} = \\ = \end{matrix} \begin{bmatrix} -1 \\ -1 \end{bmatrix}$$

$$C \quad \begin{bmatrix} 20-12 \\ 20-12 \end{bmatrix} \begin{matrix} = \\ = \end{matrix} \begin{bmatrix} +8 \\ +8 \end{bmatrix}$$

Figure 4-2. Observed difference between weighted group means and total mean.

You will note in Figure 4-2 that the group mean differences are either positive or negative deviations from the total mean. The sum of these deviations provides us with no information concerning the difference of the group means because this sum is equal to zero. If we square each deviation, we can eliminate the negative sign; and furthermore, a summing of these squared values gives us a value which represents the magnitude of the deviations squared. This sum of squares (SS_a) divided by the number of subjects (N) from which the sum of squares were calculated is the among group variance $\left((SS_a) / N = \text{Among Variance} = SD_a^2\right)$ or the observed variance of treatment means. (See Figure 4-3 for this calculation.)

Treatment	Subject	Treatment Mean ($\bar{X}$)	Total Mean (M)	$\bar{X}$-M	$(\bar{X}\text{-M})^2$
A	1	5	12	-7	49
	2	5	12	-7	49
B	3	11	12	-1	1
	4	11	12	-1	1
C	5	20	12	+8	64
	6	20	12	+8	64

The sum of $(\bar{X}\text{-M})^2 = 228$ is called "sum of squares among groups" (SS_a). The observed variance of treatment means is $(SS_a)/N = 38$ (often labeled SD_a^2).

Figure 4-3. Calculation of observed variance of treatment means.

Expected Chance Variance of Treatment Means. The expected chance variance of the treatment means is estimated by calculating the variation of the subject's observed score from the mean of his group. Figure 4-4 shows the calculation of this within group variance.

Treatment	Subject	(X-$\bar{X}$)		(X-$\bar{X}$)	$(X-\bar{X})^2$
A	1	4-5	=	-1	1
	2	6-5	=	+1	1
B	3	10-11	=	-1	1
	4	12-11	=	+1	1
C	5	20-20	=	0	0
	6	20-20	=	0	0

The sum of $(X-\bar{X})^2 = 4$ is called "sum of squares within groups" (SS_w). The within group variance is simply $(SS_w)/N = 4/6 = SD_w^2 = 2/3 = .667$.

Figure 4-4. Calculation of within group variance.

As can be seen in Figure 4-4, the scores within each group are very close to the mean of the group. When the variance among groups is large and the variance within groups is small, the F value becomes larger. A large F value indicates that the observed mean differences may not be due to just chance sampling fluctuations. Before we calculate the F, we must discuss the notion of degrees of freedom.

Degrees of Freedom. Blommers and Lindquist (1960) have defined the degrees of freedom of a given statistic as ". . . the number of observations involved minus the number of necessary auxiliary values which are themselves derived from the observations."

An example of this definition will be shown with respect to the example found in Figures 4-2 and 4-4. In Figure 4-2 the degrees of freedom are found in the following manner:

1. We are trying to find the variability that is observed among the three treatment means. Therefore, we will be dealing with three observations.
2. In determining the variance of the three values, we must determine one auxiliary value -- the mean of the three values (note that this mean is dependent on only the three observed values).
3. The degrees of freedom (2) is equal to the number of observations (3) minus the number of auxiliary values (1).

In Figure 4-4 the degrees of freedom are different from the degrees of freedom (df) in Figure 4-2. The df in Figure 4-4 are found in the following manner:

1. We are concerned with the variance in each treatment sample. Therefore, we are concerned with the variability of each score from the respective treatment means. There are six observed scores.
2. In determining the variance for each treatment sample, we must find the mean of each sample -- in this example there are three treatment means to be determined; therefore, we have three auxiliary values.
3. The degrees of freedom (3) is equal to the number of observations (6) minus the number of auxiliary values (3).

The beginning student may experience some difficulty with respect to this definition. Therefore, we shall provide general formulas when needed in this text. The general formulas provided in this text are based on the above definition of degrees of freedom.

The F Ratio. Given the SD_a^2, SD_w^2, degrees of freedom (df_1) numerator, and df_2 denominator, we can calculate the F ratio.

$$F = \frac{(SD_a^2) / df_1}{(SD_w^2) / df_2}$$

where:

$SD_a^2 = 38$; $df_1 = 2$

$SD_w^2 = 2/3 = .667$; $df_2 = 3$

$$F = \frac{(38/2)}{(.667/3)} = \frac{19.0}{.2225} = 85.5$$

Since the distribution of F changes when the degrees of freedom associated with the F ratio change, the degrees of freedom must be used to determine the probability of obtaining observed mean differences as large or larger due to chance. If we enter an F table (Guilford, 1965, p. 583) with 2 and 3 degrees of freedom, we find that an F of 30.82 or larger theoretically would be encountered in the F-distribution (df = 2, 3) less than one time in a hundred. The probability of encountering in the F-distribution an F of 85.5 (with 2 and 3 degrees of freedom) is highly improbable. Therefore, we can conclude with some confidence that the means, 5, 11, and 20 are not the same. Apparently Treatment C is most effective, Treatment B next best, and Treatment A least effective in leading to success on the criterion test.

The reader familiar with tables reporting analysis of variance may not recognize the formula used to calculate F. Usually a summary table such as Figure 4-4a is reported in the journals.

Source	SS	df	MS	F
Between (among)	228	2	114	85.5
Within	4	3	1.33	
total	232	5		

The data are from Example I.

Figure 4-4a. A typical table used to report analysis of variance.

You should note the sums of squares (SS), which we calculated on page 48, are usually reported rather than the variances. Since dividing 228 and 4 by N does not change the ratio, this added calculation is not reported in the literature.

Please note: Although means of such a difference in magnitude are unlikely when the within variance is so small, there is always the possibility that this particular sample is that rare occasion. As was mentioned in Chapter II, we set the point of probability from which we shall reject the no-difference hypothesis and live with it. Many factors enter into the amount of risk the investigator is willing to tolerate. Furthermore, even when the difference between (among) means is statistically significant, a cost payoff decision might be considered. For example, even though one treatment adds five points to a group mean, the cost of the treatment might not be worth the payoff. Conversely, if a treatment facilitates group performance but the statistical difference might be less than set originally (i.e., $p < .05$ instead of $p < .01$), the investigator might still recommend the marginal treatment if it cost the same as the treatments which are of questionable inferiority. These practical considerations will be revisited.

Traditional Analysis of Variance: Example II

In order to show how the F ratio is influenced by the within group variation, let us consider another example which might have been obtained from the three treatment group examples. In this example, let us say the scores on the physics test for each group were: A (2, 8); B (6, 16); and C (15, 25). The mean for Treatment A = 5, Treatment B = 11, and Treatment C = 20. These means are identical to those reported in Figure 4-1, and the SS_a would be 228 and the among group variance equals 38, also identical to the reported variance in Figure 4-2. (See Figure 4-2. Can you see the identical elements?) The within group variance for Example II is shown in Figure 4-5.

		$X-\bar{X}$		$X-\bar{X}$	$(X-\bar{X})^2$
A	2	2-5	=	-3	9
	8	8-5		+3	9
B	6	6-11	=	-5	25
	16	16-11		+5	25
C	15	15-20	=	-5	25
	25	25-20		+5	25

Mean A = 5, B = 11, and C = 20. 118 = sums of squares; therefore, $SD_w^2 = 118/6 = 19.66$.

Figure 4-5. Calculation of the within group variance for Example II.

If we substitute the new SD_W^2 in the equation for calculation of the F ratio, we get:

$$F = \frac{38/2}{19.66/3} = \frac{19.0}{6.55} = 2.8$$

An examination of an F table shows that an F ratio of at least 2.8 with 2 and 3 degrees of freedom occurs much more often than five times in a hundred. Therefore, we can conclude (again with some confidence) that the means 5, 11, and 20 are not statistically different. Apparently the treatments do not differ significantly in effectiveness.

This overview of the analysis of variance was provided to give some notion of how mean group differences are analyzed to determine the likelihood that sampling error can account for the observed differences. The analysis depends upon (1) group mean differences from the total mean, and (2) the variation of individual scores within the group. Essentially, as was shown in Example II, if within group variation is great, then the mean differences must be much greater to be a highly unlikely observation in the F-distribution.

Conversely, when within group variation is small, an observed among group mean difference of only moderate value is necessary to be unlikely and, therefore, significant.

A more complete discussion regarding the F-statistic can be found in Hays (1963).

Multiple Regression Analysis: Example I

Example I, regarding the influence of three programs upon performance on a physics test, can be examined using a predictive equation. The following discussion will present the same example from this approach.

We might phrase the research question: "Are physics programs A, B, and C equally effective in influencing performance on the physics test?"

Essentially we are asking: "Can knowledge of treatment enable us to predict criterion test scores with fewer errors?" Using the information presented in Chapter III regarding linear combinations of vectors, we can cast a set of vectors, place restrictions upon the set, and note the influence the restrictions have upon error.

In dealing with the data in Example I, we can form a vector Y_1 of dimension 6 to represent the six physics test scores:

$$Y_1 = \begin{bmatrix} Y_{1_1} \\ Y_{1_2} \\ Y_{1_3} \\ Y_{1_4} \\ Y_{1_5} \\ Y_{1_6} \end{bmatrix} = \begin{bmatrix} 4 \\ 6 \\ 10 \\ 12 \\ 20 \\ 20 \end{bmatrix}$$

where:

Y_{1_1} = physics test score for individual 1 = 4

Y_{1_2} = physics test score for individual 2 = 6

Y_{1_3} = physics test score for individual 3 = 10

Y_{1_4} = physics test score for individual 4 = 12

Y_{1_5} = physics test score for individual 5 = 20

Y_{1_6} = physics test score for individual 6 = 20

Furthermore, we can generate three vectors, each of dimension 6 to represent each of the three treatment conditions:

X_1 = a vector in which the element is 1 if the corresponding score on Y_1 is from a subject who had Program A; zero otherwise.

X_2 = a vector in which the element is 1 if the corresponding score on Y_1 is from a subject who had Program B; zero otherwise.

X_3 = a vector in which the element is 1 if the corresponding score on Y_1 is from a subject who had Program C; zero otherwise.

These three vectors would look like this:

$$X_1 = \begin{bmatrix} 1 \\ 1 \\ 0 \\ 0 \\ 0 \\ 0 \end{bmatrix} \qquad X_2 = \begin{bmatrix} 0 \\ 0 \\ 1 \\ 1 \\ 0 \\ 0 \end{bmatrix} \qquad X_3 = \begin{bmatrix} 0 \\ 0 \\ 0 \\ 0 \\ 1 \\ 1 \end{bmatrix}$$

You will note that no individual has been given both treatments (this is often called a case of mutually exclusive categories). The following discussion will hold for all cases where mutually exclusive categories are used as long as the criterion values meet the assumptions delineated earlier.

We may now wish to raise the question: "Is the vector Y_1 some weighted linear combination of X_1, X_2, and X_3?" or

$$Y_1 = a_1X_1 + a_2X_2 + a_3X_3?$$

where:

a_1, a_2, and a_3 are unknown weights which make the linear statement true.

Let us form the four vectors using the data from Example I:

$$\begin{bmatrix} 4 \\ 6 \\ 10 \\ 12 \\ 20 \\ 20 \end{bmatrix} = a_1 \begin{bmatrix} 1 \\ 1 \\ 0 \\ 0 \\ 0 \\ 0 \end{bmatrix} + a_2 \begin{bmatrix} 0 \\ 0 \\ 1 \\ 1 \\ 0 \\ 0 \end{bmatrix} + a_3 \begin{bmatrix} 0 \\ 0 \\ 0 \\ 0 \\ 1 \\ 1 \end{bmatrix}$$

Since individual 1 and individual 2 did not obtain equal scores on Y_1, there is no a_1 which will contribute to making the linear combination true. This holds also for a_2, but since both Treatment C individuals performed the same, a_3 could add to the equality if a_1 and a_2 were possible values. If weights were available to make $Y_1 = a_1X_1 + a_2X_2 + a_3X_3$ true, then linear dependency would exist and perfect prediction could be obtained using knowledge of group membership.

Lacking perfect prediction, we can produce a new vector (E_1) which, when combined with the weighted three groups' membership vectors, will make vector Y_1 a linear combination of the four vectors:

$$Y_1 = a_1X_1 + a_2X_2 + a_3X_3 + E_1$$

In order to satisfy the equality expression, the elements in vector E_1 must be the difference between the observed score (Y_{1_1}, Y_{1_2}, . . . Y_{1_6}) and the weight associated with the individual's group. An infinite number of weights are possible; but if we desire to minimize the squared values in vector E_1, the criterion mean for the particular group is best. Consider Example I:

$$\underset{Y_1}{\begin{bmatrix} 4 \\ 6 \\ 10 \\ 12 \\ 20 \\ 20 \end{bmatrix}} = 5\underset{X_1}{\begin{bmatrix} 1 \\ 1 \\ 0 \\ 0 \\ 0 \\ 0 \end{bmatrix}} + 11\underset{X_2}{\begin{bmatrix} 0 \\ 0 \\ 1 \\ 1 \\ 0 \\ 0 \end{bmatrix}} + 20\underset{X_3}{\begin{bmatrix} 0 \\ 0 \\ 0 \\ 0 \\ 1 \\ 1 \end{bmatrix}} + \underset{E_1}{\begin{bmatrix} -1 \\ 1 \\ -1 \\ 1 \\ 0 \\ 0 \end{bmatrix}}$$

a_1 = 5 since the two scores obtained by the members of this treatment group were 4 and 6.

$$a_2 = 11 = \frac{10 + 12}{2}$$

$$a_3 = \frac{20 + 20}{2}$$

The first element in vector E_1 is -1, which represents the difference of Subject One's score (4) from the weight assigned to his treatment group (5). The element in vector E_1 associated with Subject Two is +1 [6 (observed score) minus 5 (weight of X_1)].

When the weights associated with X_1, X_2, and X_3 are the appropriate means, an interesting fact about vector E_1 should be noted. If we square each element in vector E_1 to remove signs, sum the squared values, and divide by N, we get the within group variance (the same as we calculated in Figure 4-4, when dealing with the analysis of variance). See Figure 4-6.

$$E_1 = \begin{bmatrix} +1 \\ -1 \\ +1 \\ -1 \\ 0 \\ 0 \end{bmatrix} \qquad \text{By squaring each element in } E_1\text{, we get a new vector } E_4 \qquad E_4 = \begin{bmatrix} 1 \\ 1 \\ 1 \\ 1 \\ 0 \\ 0 \end{bmatrix}$$

The sum of the elements of this new vector $E_4 = 4$ is called "sums of squares within group" (SS_w). The within group variance = SD_w^2 = $4/6 = 2/3 = .667$.

Figure 4-6

Our question regarding ". . . are the three programs equally good?" can be expressed as in our linear equation as: $a_1 = a_2 = a_3 = a_0$ (the three weights are equal to a_0, which is a weight common to all treatments). This can be expressed:

$$Y_1 = a_0X_1 + a_0X_2 + a_0X_3 + E_2$$

You should note that E_2 is a new error vector whose elements now will be the difference of the individual's observed score and some weight common to all groups. Figure 4-7 shows the elements in the five vectors and the common weight. Note that 12 was used as the common weight since it is the value which minimizes the squared error.

$$\underset{Y_1}{\begin{bmatrix} 4 \\ 6 \\ 10 \\ 12 \\ 20 \\ 30 \end{bmatrix}} = 12 \underset{X_1}{\begin{bmatrix} 1 \\ 1 \\ 0 \\ 0 \\ 0 \\ 0 \end{bmatrix}} + 12 \underset{X_2}{\begin{bmatrix} 0 \\ 0 \\ 1 \\ 1 \\ 0 \\ 0 \end{bmatrix}} + 12 \underset{X_3}{\begin{bmatrix} 0 \\ 0 \\ 0 \\ 0 \\ 1 \\ 1 \end{bmatrix}} + \underset{E_2}{\begin{bmatrix} -8 \\ -6 \\ -2 \\ 0 \\ +8 \\ +8 \end{bmatrix}}$$

E_2 was calculated where $a_0 = 12$, and each element in E_2 represents the difference of the common weight from the individual's observed score.

Figure 4-7

Since a_0, a common weight, is applied to each group and since the elements in vectors X_1, X_2, and X_3 are ones and zeros and each individual is represented only once, we can cast a new vector which is a linear combination of X_1, X_2, and X_3. We can call this the unit vector U. (See Chapter III.)

$$\underset{U}{\begin{bmatrix} 1 \\ 1 \\ 1 \\ 1 \\ 1 \\ 1 \end{bmatrix}} = \underset{X_1}{\begin{bmatrix} 1 \\ 1 \\ 0 \\ 0 \\ 0 \\ 0 \end{bmatrix}} + \underset{X_2}{\begin{bmatrix} 0 \\ 0 \\ 1 \\ 1 \\ 0 \\ 0 \end{bmatrix}} + \underset{X_3}{\begin{bmatrix} 0 \\ 0 \\ 0 \\ 0 \\ 1 \\ 1 \end{bmatrix}}$$

The unit vector is a linear combination of X_1, X_2, and X_3. Therefore, $a_0U = a_0X_1 + a_0X_2 + a_0X_3$.

Figure 4-8

If we substitute the identity $a_0U = a_0X_1 + a_0X_2 + a_0X_3$ into the linear equation $Y_1 = a_0X_1 + a_0X_2 + a_0X_3 + E_2$, we get: $Y_1 = a_0U + E_2$. Figure 4-9 shows the elements in these new vectors.

$$\underset{Y_1}{\begin{bmatrix} 4 \\ 6 \\ 10 \\ 12 \\ 20 \\ 20 \end{bmatrix}} = 12 \underset{U}{\begin{bmatrix} 1 \\ 1 \\ 1 \\ 1 \\ 1 \\ 1 \end{bmatrix}} + \underset{E_2}{\begin{bmatrix} -8 \\ -6 \\ -2 \\ 0 \\ +8 \\ +8 \end{bmatrix}}$$

Note the elements in E_2 are identical to those represented in Figure 4-7.

Figure 4-9

Furthermore, if we square each element in E_2, sum these values, and divide by N, we get a value which is called the total variance of Y_1. See Figure 4-10.

$$\underset{E_2}{\begin{bmatrix} -8 \\ -6 \\ -2 \\ 0 \\ +8 \\ +8 \end{bmatrix}} \quad \underset{E_3}{\begin{bmatrix} 64 \\ 36 \\ 4 \\ 0 \\ 64 \\ 64 \end{bmatrix}}$$

The elements of E_3 are the squared values of the corresponding elements of E_2

The sum of the squared elements in $E_2 = 232$.
$232 =$ sum of squares total (SS_t). This value, divided by N, is equal to SD_t^2 (total variance). $232/6 = 38.667 = SD_t^2$.

Figure 4-10

The Components in the Analysis of Variance

The squared deviations of the scores from the total mean is called the total variance (SD_t^2). Look at Figure 4-3. The among group variance (SD_a^2) was 38 (228/6). In Figure 4-4, the within group variance $SD_w^2 = .667$. If we sum these two values, lo and behold, we get 38.667, which is the total variance. Indeed this is what analysis of variance is all about. We want to determine whether knowledge of treatments, etc., can account for more than a chance amount of the total variation.

The F-Test Using Regression Values

The F ratio was calculated using the following formula:

$$F = \frac{SD_a^2 / df_1}{SD_w^2 / df_2}$$

Since SD_w^2 was obtained from operations upon the vector E_1 and the total variance (SD_t^2) was derived using operations upon vector E_2, we can calculate the F ratio because $SD_a^2 = SD_t^2 - SD_w^2$. Therefore:

$$F = \frac{(38.667 - 0.667) / df_1}{.667 / df_2}$$

The degrees of freedom are:

$df_1 = m_1 - m_2$

$df_2 = N - m_1$

where:

m_1 = the number of auxiliary values used to obtain SD_w^2. In linear regression terms, m_1 is the number of unknown weights used in the linear model from which SD_w^2 was derived. Three group means were calculated. Therefore, $m_1 = 3$.

m_2 = the number of unknown weights (auxiliary values) used to derive SD_t^2. With the unit vector in Example I, a_0 was the only weight calculated. Therefore, $m_2 = 1$.

$df_1 = 3-1 = 2.$

$df_2 = 6-3 = 3.$

Substituting these values in the F ratio, we get:

$$F = \frac{(38.667 - 0.667) / 2}{0.667 / 3} = \frac{19.0}{.2225} = 85.5$$

Note these terms are identical with those derived using traditional analysis of variance. Because we have the same F and associated degrees of freedom, we come to the same conclusion that it is unlikely for an F of this magnitude to occur; thus, the three treatments are probably <u>not</u> equal.

<u>Summary of Example I</u>

The method of selecting weights so as to minimize the sum of the squares of the elements in the error vector is called the "least squares" procedure. The comparison of the variance (SD_w^2) obtained with knowledge of some sort (here we knew which treatment each individual received) to the variance obtained without the knowledge of group membership (SD_t^2) provides an estimate of the value of the knowledge. Multiple regression analysis provides a least squares solution for obtaining the various weights. This procedure is more fully discussed in Hays (1963).

<u>Review</u>

1. We had information regarding performance on a physics test. These values were cast into a vector Y_1. The dimension of the vector was 6.

2. Information regarding group membership was given and with this knowledge we generated three new vectors (X_1, X_2, and X_3), each of dimension 6. The elements were ones (1) and zeros (0).
3. Weights for vectors X_1, X_2, and X_3 were calculated so as to make the sum of squares of the elements in a vector E_1 a minimum. These weights were $a_1 = 5$; $a_2 = 11$; and $a_3 = 20$.
4. The sum of the squared elements in vector E_1 divided by N was shown to be the within group variance (SD_w^2).
5. Since our research question asked ". . . are the effects of Treatments A, B, and C equal?", the weights a_1, a_2, and a_3 were set equal to a common weight, a_0. In this special case a_0 was the overall mean (grand mean) = 12.
6. Vector E_2 with the elements squared and summed, divided by N gave total variance (SD_t^2).
7. Among group variance was shown to be $(SD_t^2 - SD_w^2) = SD_a^2$.

The Relationship of Sums of Squares Calculated by Least Squares Method to R^2

.We now need to consider another aspect of the multiple regression analysis that we will use in the future for calculating the F ratio -- that is, R^2. R^2 can best be explained by returning for a moment to our discussion of sums of squares, involving the physics exam problem, Example I. In that example, we calculated (Figure 4-3) the observed variance of treatment means or $(SS_a)/N$, where SS_a is the sum of squares among groups, the within group variance (Figure 4-4) or $(SS_w)/N$, and then calculated the F ratio using those values (page 51).

We could add SS_a and SS_w to get the total sums of squares (SS_t) and divide by N to get the total variance (total variance = $\frac{SS_a + SS_w}{N}$ = SS_t/N). We could also calculate the SS_t directly from the data. (It should be noted here that the sum of squares is also referred to as error sums of squares (ESS).)

Consider the scores on the physics exam:

$$Y_1 = \begin{bmatrix} 4 \\ 6 \\ 10 \\ 12 \\ 20 \\ 20 \end{bmatrix}$$

We can determine the mean of these scores and construct an error vector which is an estimate of the dispersion of scores from the mean. In the case presented above, the mean is 12 and the error vector (observed score minus the mean) is:

$$E_1 = \begin{bmatrix} -8 \\ -6 \\ -2 \\ 0 \\ +8 \\ +8 \end{bmatrix} \xrightarrow[\text{to remove signs}]{\text{squaring each element}} E_2 = \begin{bmatrix} 64 \\ 36 \\ 4 \\ 0 \\ 64 \\ 64 \end{bmatrix}$$

Summing the scores in E_2, the sum of the squared errors equals 232. For purposes of an analysis of variance, this sum is called the "total sums of squares" (SS_t). (We now have the value that we could have obtained by adding our SS_a, which was 228, and our SS_w, which was 4 -- from Figures 4-3 and 4-4.)

Suppose that we were to start with our data (the test scores and our knowledge of group membership) and calculate:

$SS_t = 232$ and $SS_w = 4$

When we use the knowledge of group membership and calculate weights for those vectors to predict the test scores, our error sum of squares is small (4). When we do not use that knowledge and, instead, predict the test scores using only a weighted unit vector, the error sum of squares is much larger (232).

To find out the probability of finding a reduction of accounted for variance of this magnitude and with these degrees of freedom, we perform an F-test.

Since $SS_a = SS_t - SS_w$, we can write the formula for F as:

$$F = \frac{(SS_t - SS_w) / (m_1 - m_2)}{SS_w / (N - m_1)}$$

where:

m_1 = the number of linearly independent vectors used to calculate SS_w.

m_2 = the number of linearly independent vectors used to calculate SS_t.

In the example presented here, $m_1 = 3$ since the three means were calculated (e.g., 5, 11, 20). $m_2 = 1$ because we calculated only one mean (12) to get SS_t.

$N = 6$.

$$F = \frac{(232 - 4) / (3 - 1)}{4 / (6 - 3)} = \frac{114}{1.13} = 85.5 \qquad df = 2/3$$

As we found when we calculated this F ratio previously, the probability of encountering a change of this magnitude when going from calculating ESS with knowledge of group membership to calculating without such knowledge is less than .01 ($p < .01$).

When we calculated F much earlier in this chapter -- under the analysis of variance -- we used:

(SS_a) / N instead of SS_a $(SS_t - SS_w)$

and

(SS_w) / N instead of SS_w.

You should note that dividing by N does not alter the ratio between the numerator and denominator; we got the same F ratio, 85.5, each time. Calculation without dividing by N is easier since it involved one less operation.

There is another calculation which will give us the same ratio and is easier still. This one involves the R^2, which we mentioned several pages back.

In large studies the calculation of an error vector and then the sums of squares using weights derived from a least squares solution can be cumbersome. The R^2, which reflects the amount of total variance in the criterion vector which is predicted by the predictor variables, can be used to calculate the F ratio.

Now we can consider the relationship of R^2 to the sums of squares that we have been using. In our example, the total variance (SD_t^2) is 38.667. Our within group variance (SD_w^2) is .667. We know that the total variance minus the within group variance equals the variance among groups (SD_a^2). This variance among groups is called the variance accounted for by using knowledge of group membership and is:

$$38.667 - .667 = 38.00$$

Then the per cent of total variance accounted for by using knowledge of group membership can be expressed:

$$\frac{\text{variance accounted for}}{\text{total variance}} = \frac{38.000}{38.667} = 98.27\%$$

or:

$$R^2_f = \frac{(SD_t^2) - (SD_w^2)}{(SD_t^2)} = \frac{38.667 - .667}{38.667} = \frac{38.000}{38.667} = .9827$$

Since this R^2 is calculated using full knowledge of group membership, we call it R^2_f. The subscript "f" refers to "full." We read this R^2_f as "R squared, full model."

If we wish to compare the per cent of variance explained by group membership with the per cent explained without knowledge of group membership, we need to calculate an R^2_r (R squared, restricted model).

In this particular example, when we have no knowledge of group membership, we have only knowledge of the mean of the combined groups (grand mean); then $(SD_w^2) = (SD_t^2)$. Therefore:

$$R^2_r = \frac{(SD_t^2) - (SD_w^2)}{(SD_t^2)} = \frac{(38.667 - 38.667)}{38.667} = \frac{0}{38.667} = .00$$

using the formula:

$$F = \frac{(R^2_f - R^2_r) / (m_1 - m_2)}{(1 - R^2_f) / (N - m_1)} = \frac{(.9827 - 0) / (3 - 1)}{(1 - .9827) / (6 - 3)} = \frac{.9827 / 2}{.0173 / 3} =$$

$$\frac{.49135}{.00576} = 85.5 \qquad df = 2/3$$

Please note the ratio equivalence of:

Numerators

$R^2_f - R^2_r$ = per cent of total variance accounted for with knowledge of group membership.

(SD_a^2) = amount of variance accounted for with knowledge of group membership.

Denominators

$(1 - R^2_f)$ = per cent of variance within groups, or per cent of unaccounted for variance.

(SD_w^2) = amount of variance within groups or unaccounted for variance.

In essence, whether we calculate the F ratio using (1) error sums of squares; (2) variances; or (3) per cent of variance (R^2) has no effect on the F ratio since the ratio among the three terms remains equivalent. We use R^2 (per cent of variance) because it is more economical to use with machines.

Assumptions Underlying the F-Statistic

For the ratio $\dfrac{(R^2_f - R^2_r)/(m_1 - m_2)}{(1 - R^2_f)/(N - m_1)}$ to be distributed as an F, several assumptions are made by the researcher about the criterion available. (Please note that these assumptions are made with respect to the criterion variable and are not assumed about the predictor variables.)

1. The sample of N observations is drawn at random.
2. The treatment populations (as defined in Chapter II) are normally distributed with respect to the criterion variable.

3. The elements in the criterion vector are independently distributed variables with a common variance (SD^2) for each of the treatment populations.

The first assumption is concerned with the manner in which elements in the criterion vector are selected. You may recall that in Chapter II the meaning of randomness was discussed at some length. Theoretically, the assumption of randomness means that each subject in the population had an equal chance of being selected in the sample. To theoretically satisfy this assumption would mean that the researcher would have to have access to the total population. In most studies this is unreasonable, and methods (previously discussed in Chapter II) are used to approximate this assumption.

The second assumption infers that the population available for this study is normally distributed with respect to _only_ the criterion used in the study. We hasten to point out to the reader that if a population is abnormal with respect to one attribute, this does _not_ mean that the population is abnormal with respect to all attributes. The reader must note that this assumption is made with respect to the population, and if the _sample_ used in the study is not normal, the assumption of normality has not necessarily been violated.

The third assumption contains two important points. One point of concern is that the variables have a common variance of (SD^2). This assumption is commonly called "homogeneity of variance" and has been discussed extensively by Winer (1962). The second point of concern in this assumption is that the elements in the criterion vector are independent. Essentially, this means that the same individuals cannot be represented more than once in the criterion vector. If the same individuals are used in the criterion vector, the third assumption is not met.

In reviewing the assumptions, it is obvious that in many cases the researcher is unable to assume that he has theoretically met all of the assumptions. Failure to meet these assumptions would mean that the ratio previously defined as F could be assumed to be distributed only approximately as an F. If all of the assumptions are met, the ratio is said to be an exact test. In an exact test, the probability of making a Type I error (see Chapter II) is exactly the level of significance. In an approximate test the proabaility of making a Type I error is not exactly known, but the researcher aften assumes that this probability is close to the level of significance.

In reviewing the above discussion, it is clear that most research studies are, at best, dealing with approximate inferential tests. Since these tests are approximate, it is imperative that we as researchers concern ourselves with the replication of research studies (cross-validation). A more thorough discussion of cross-validation is presented in Chapter VI of this text.

Problems for Chapter IV

1. The criterion test scores in Example II can be cast into a vector Y_1 and can be shown to be a linear combination of $a_1X_1 + a_2X_2 + a_3X_3 + E_3$. Express these five vectors in the form of the vectors on the bottom of page 57.

2. (a) Obtain a table of F and find the F value necessary to occur one time in a hundred (the .01 point) for the following degrees of freedom:

df numerator	df denominator	F
1	100	____
2	100	____
3	100	____
5	200	____
5	300	____
5	400	____
500	1000	____

(b) What seems to happen when the df in the numerator increases as the df in the denominator remains constant?

(c) What seems to happen when the df in the denominator increases as the df in the numerator remains constant?

3. You have randomly assigned from a class of 24 subjects eight subjects to one of three test conditions. Under Condition 1 the test given was said to be very important and would account for 10 per cent of the course grade. Under Condition 2 the test result was not to be counted for course grade. Under Condition 3 the test result would account for 90 per cent of the course grade.

Your research question is: "Does the importance of the test condition influence test performance?"

Students 1 through 8 were in Condition 1 and received the following scores on the test: 3, 5, 2, 4, 8, 4, 3, 9.

Students 9 through 16 were in test Condition 2 and scored: 4, 4, 3, 8, 7, 4, 2, 5.

Condition 3 students 17 through 24 scored: 6, 7, 8, 6, 7, 9, 10, 9.

(a) Set up an equation where the criterion vector Y_1 is a linear combination of the weighted vectors: $X_1 = 1$ if Condition 1; $X_2 = 1$ if Condition 2; and $X_3 = 1$ if Condition 3; and E_5. Please note: This asks for the equation -- not the expression of vectors.

(b) Now express the five vectors for this equation in the form on page 57.

(c) Now try the tough problem -- try to answer the research question by obtaining the $SD_w{}^2$ and $SD_t{}^2$ and performing an F-test. This may be asking too much, but please try!

Chapter V Organization of Program Linear

We have only begun to discuss the use of multiple linear regression analysis, but we have looked rather closely at one example of a question you might ask: "Are physics programs A, B, and C equally effective in influencing performance on the physics test?" We discussed how we would go about answering this question by analysis of variance and by regression analysis, pointing out why we prefer regression analysis. This chapter is designed to show you how to perform this regression analysis using a computer program. (We call this "Program LINEAR.") We shall use the physics exam problem #1 from Chapter IV to prepare you to run the LINEAR program on the computer.

The program that actually performs the regression analysis is presented in the Appendix. For now, all you need to know about the program is the group of cards that you put with your data cards and submit to the computer. These cards essentially call out the rest of the program from the computer.

Figure 5-1 shows a list of the cards in the program deck.

```
$IBSYS
$JOB                7040     70 241186 50206 3 JOHN DOE
$IBJOB          NODECK
$IBFTC DATRAN   NODECK
      SUBROUTINE DATRAN(I,NFRCOL,J,NTOCOL)
C     DATRAN                7 JANUARY 1964            PAGE 1
C     TRANSFORMS DATA MOVING FROM TPTOTP
      DIMENSION FMT1(22), KFMT1(22),FMT2(22),KFMT2(22),A(1),KA(1)
      COMMON FMT1,FMT2,A
      EQUIVALENCE (FMT1,KFMT1),(FMT2,KFMT2),(A,KA)
      CALL ZEROST (J+1,1,NTOCOL)
   10 CALL MOVCOR(I+1,1,NFRCOL,J+1)
   20 RETURN
   30 END
$IBFTC LINEAR NODECK
      PROGRAM LINEAR WITH TIMING AND USE OF ONE DISC UNIT
      DIMENSION FMT1(22),FMT2(22),KFMT1(22),KFMT2(22),A(11000),KA(11000)
      COMMON FMT1,FMT2,A,IT(1)
      EQUIVALENCE (FMT1,KFMT1),(FMT2,KFMT2),(A,KA)
    1 CALL TAPGEN(8)
      ITIME = IT(8)
      READ (5,5) (FMT1(1),I=1,12)
      IF (FMT1(1) - FMT1(2)) 10, 15, 10
    5 FORMAT (12A6)
   10 WRITE (6,20) (FMT1(I),I=1,12)
   20 FORMAT (1H1,5X,12A6////)
      READ (5,25) NC,NIV,NTV
   25 FORMAT (16I5)
      CALL DATMOV(5,0,2,NC,NIV,1,8,12,12H DATA        ,10,NTV,101)
      CALL CORRLB (8,12H DATA       ,NC,NTV,1,101,201)
      CALL ZEROST(10401,1,100)
      CALL REGRED(1,101,201,10201,10301,10401,NTV)
      TM = IT(8) - ITIME
      TM = TM/60.
      WRITE(6,30) TM
   30 FORMAT(22H-TIME FOR THIS RUN IS ,F10.2,8H SECONDS )
      GO TO 1
   15 CALL EXIT
      END
$ENTRY          LINEAR
```

Figure 5-1. Listing of program LINEAR

The first four cards are systems cards that we use on the IBM-7040 computer. These will vary in other computer installations. You will need to check with your computer center to find out the procedure for using and submitting the LINEAR program.

The first problem we discuss is the physics exam problem. We show in detail how you prepare it to run on the computer. This includes the way in which you arrange information about the subjects on data cards and the way in which you prepare other cards to tell the computer how to handle those data cards. It should be emphasized that you first decide what you want to know (your research question), and then you prepare the cards that enable the computer program to aid you in answering your statistical question.

Figure 5-2 illustrates the general order of the cards. The figure will become clearer after working through several examples. Therefore, we will go through two examples with you step-by-step.

After the collection of data is completed, a data card or set of cards for each subject in the study must be prepared and punched. Similar information for each subject must be punched in the same position on such data cards for each subject. For this particular problem (the physics problem) the data could be organized in the following manner for each subject:

Column	Information Punched
1	Subject's ID number.
2	Space.
3	1 if subject is a member of Group A; and 0 otherwise.
4	1 if subject is a member of Group B; and 0 otherwise.
5	1 if subject is a member of Group C; and 0 otherwise.
6	Space.
7-8	Subject's score on physics exam.

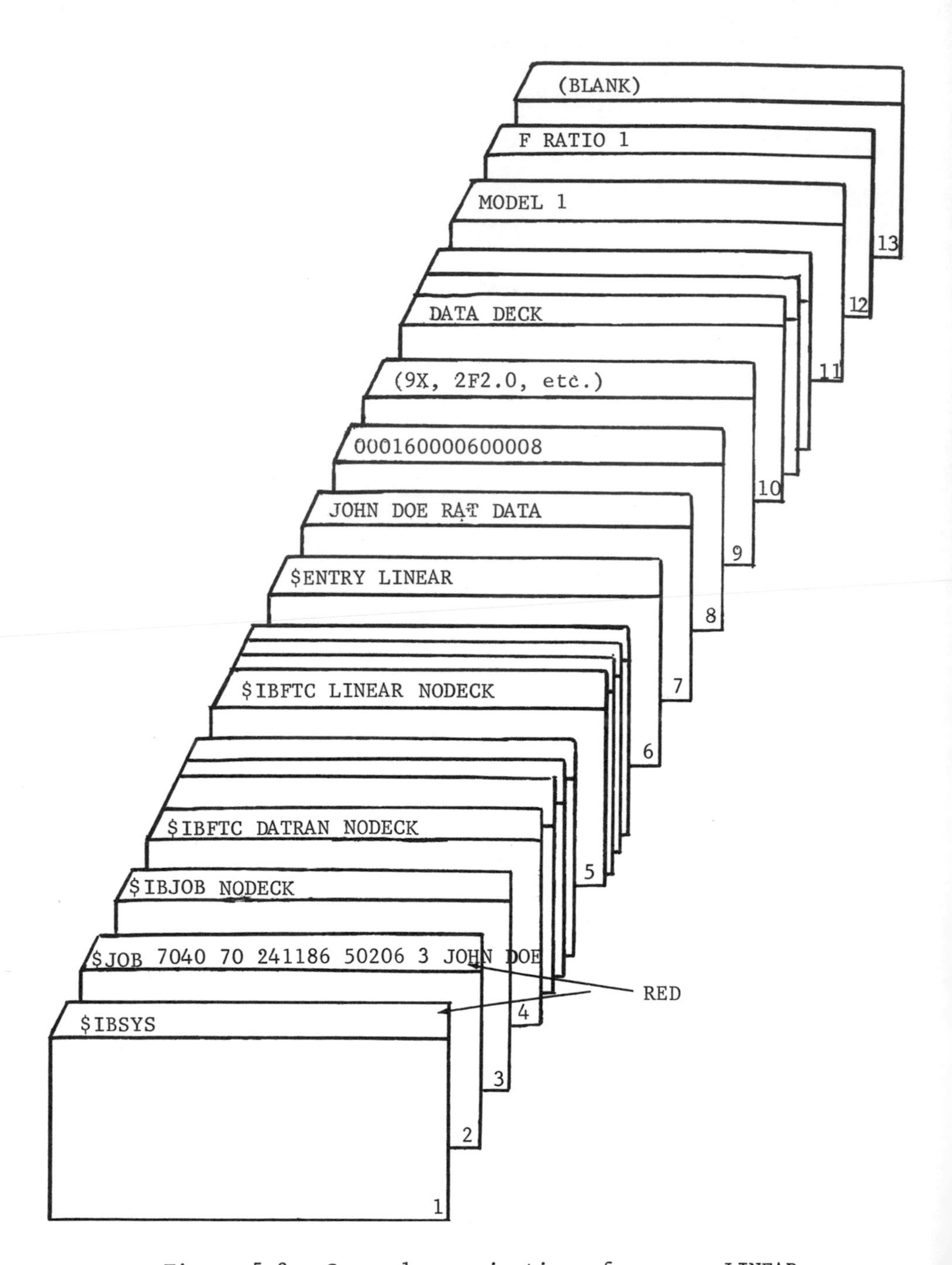

Figure 5-2. General organization of program LINEAR

It is not necessary to leave blank spaces on the data cards, but spaces are skipped here so that the data card (pictured below) can be more easily read.

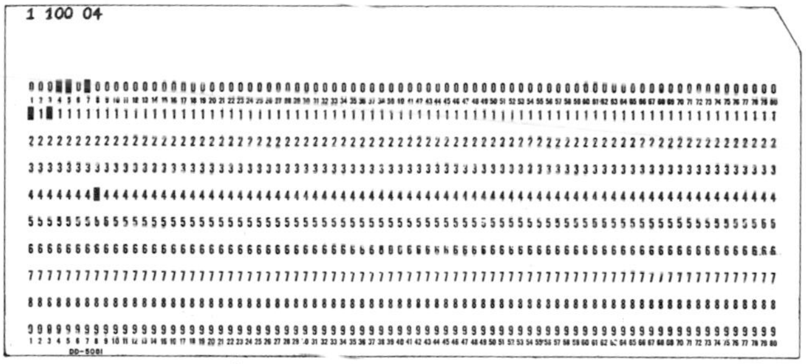

1 100 04

The card above is a general purpose IBM card. You will be using cards of this sort for the cards you punch. Notice that there are 80 numbers across the top of the card. These indicate spaces called "columns." You can put one number, letter, or symbol in each column. When an item of information takes more than one column (such as the score on the physics exam -- columns 7-8), that amount of space is called a "field." The physics exam score is, therefore, in a two-digit field.

The "1" in the first column indicates that the data card contains information about subject number 1. The "1" in the third column specified that this subject is a member of Group A, and thus cannot be a member of either Group B or Group C; so there are zeros in the fourth and fifth columns. In columns seven and eight, "04" is punched, meaning that subject number 1 obtained a score of 4 on the physics exam. It would be worthwhile for you to figure out how the cards would look for the other five subjects. (You can check these in Figure 5-3.) Cards 1-6 in Figure 5-2 remain essentially the same for all LINEAR runs. Now we will examine the cards necessary to run this particular program in the computer. These cards must be punched in a particular way. The general form of each will always be the same; the specific information will change, depending on your question and your data.

Title Card (Card #7)

A title or identification card is necessary with this program. On this card you are free to punch any information that will identify the study which is being run. The only requirement is that there must be a punch somewhere in the first twelve columns. The information on the title card will be printed on the output, so it is a good idea for your own future reference to place the number of subjects in the study on this card. For example, a title card for the physics exam problem could be:

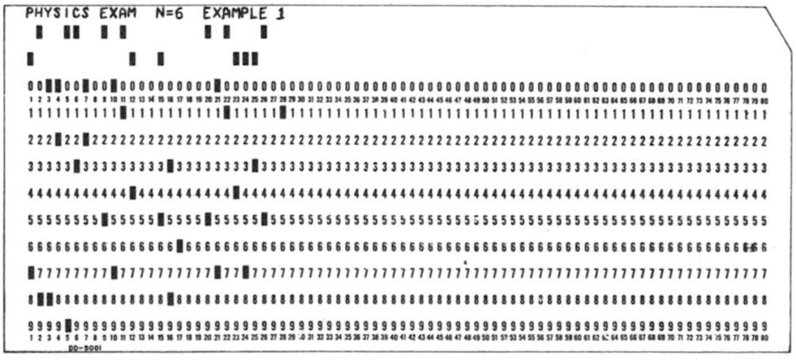

```
PHYSICS EXAM  N=6  EXAMPLE 1
```

Parameter Card (Card #8)

Indicated on the parameter card are (1) the number of subjects in the study; (2) the number of original variables read off the data cards; and (3) the total number of variables to be used in the analysis. This last value, "total number of variables," will become more meaningful as we get into more complex aspects of this method.

The specific organization of the parameter card is as follows:

Columns	Information Punched
1-5	Number of subjects in the study.
6-10	Number of original variables (read off data cards).
11-15	Total number of variables.

Pictured below is the parameter card for the physics exam problem:

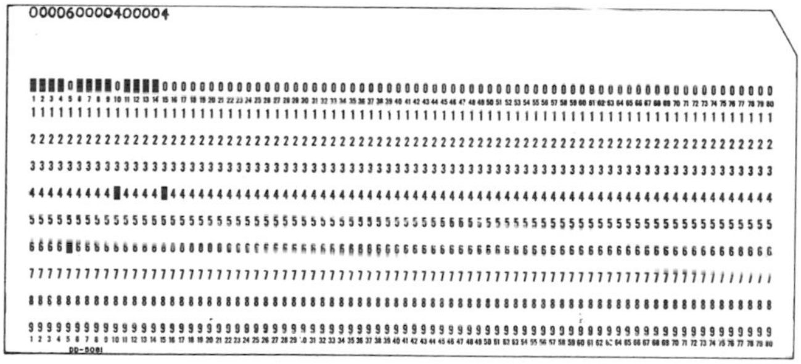

As you can see, for each of the five columns which go together, the number is placed in the right-hand position; so 6 is in column five. If the 6 were in column four, it would mean that there were 60 subjects in the study (00060). The same thing applies for columns 6-10 and 11-15.

The information on the parameter card must be punched in the prescribed positions since the program is designed for the computer to read the information in this way.

Format Card (Card #9)

In order for the computer program to organize and read correctly the information on the data cards, a format card (which tells the computer how to read the data cards) must be made. Some symbols which have special meanings in format statements must be defined before you can do this.

"X" means to skip a column without reading it, even though there may be something punched in that column. (For instance, we have a subject number in our data cards for identification, but we do not want to use it in our analysis. So we will include in our format statement instructions to skip that column.) A number, such as 10, appearing to the left of the "X" (10X) would indicate that 10 columns are to be skipped.

"F" indicates a floating point format. The general form of this format statement is: aFb.c -- where a, b, and c are whole numbers and

b = the number of columns which are to be read for this particular variable.

c = the number of decimal places for the variable.

a = the number of times this same process is to be done.

For example, 2F3.1 would mean that three columns are read together with one decimal place in the variable; and then the next three columns are to be treated the same way (the "2" indicates this). Some example of this are:

With Format	Information Printed On Data Card	Read by the Computer as
	803	80.3
F3.1	8.3	8.3
	083	8.3

As you can see, it is allowed but not necessary to punch the decimal point on the data cards. Another thing to remember is that values for both b and c must be present in the floating point format with a period separating them; if a is "1," it does not have to appear. This format is generally used since it allows either whole numbers (integers) or decimals to be read. In fact, the internal workings of the LINEAR program require data to be read by the computer in this manner.

The format card must be punched according to some definite rules: (1) The first punch (left parenthesis) must be in one of the first twelve spaces on the card; (2) Each of the elements in the statement is separated by a comma; (3) The last punch must be a right parenthesis following the last element in the statement; and (4) You may read up to 99 variables with a format card. The program will not handle more than 99 at one time. (The number of variables which LINEAR can handle may be increased by increasing the dimension of K and KA, a task that should be left to an expert programmer.) These rather simple rules become almost second-nature after some experience with the program.

We begin the format card with a left parenthesis, usually in column one. Then we put in order the directions to the computer in the form of aX or aFb.c, separating each with a comma. You may skip spaces after the comma or not, as you wish. For the physics problem, the format card could be:

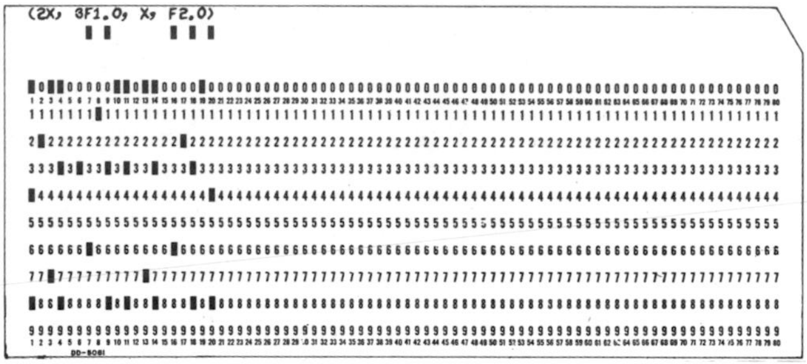

(2X, 3F1.0, X, F2.0)

This format statement indicates that the first two columns on the data card are not to be read (2X). Columns 3 through 5 are read one column at a time (3F1.0) to make up the three group membership variables. Then, the next column is skipped (X). The following two columns are read together with no decimal places (F2.0). This variable represents the score earned on the physics exam. If there is no number to the left of either the X or F, "1" is understood and the process called for is done once.

The cards that we have discussed up to this point are arranged in the following order:

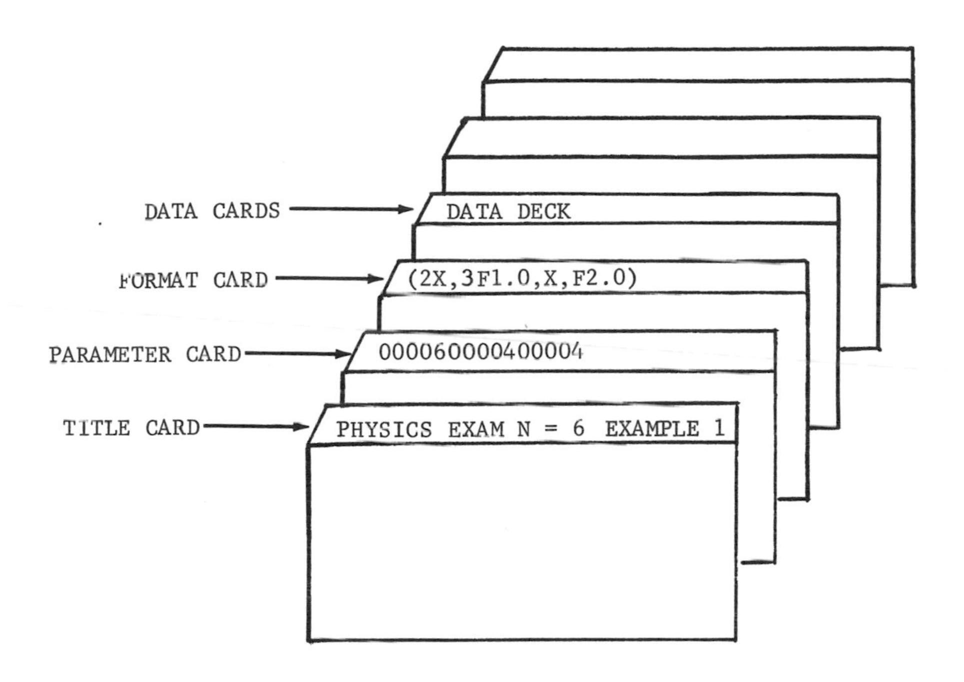

Now the program and the data are prepared, and we are ready to return to the research question raised in Chapter IV and show what further cards must be prepared to answer that question. The question was: "Are the effects of the three different programed physics texts the same as measured by the physics exam?" The full model for testing this question, as indicated in Chapter IV, is:

$$Y_1 = a_1X_1 + a_2X_2 + a_3X_3 + E_1$$

Model Card (Card #11 in Figure 5-2)

In order for the above model to be utilized by the computer, a model card must be punched and entered into the computer. The model card is set up as follows:

Column	Information Punched
1-10	Identification (e.g., Model 1)
11-13	Number of two-digit fields starting in column 22.
14-21	Stop criterion (usually 001.E-05).
22-	Specification of predictor variables and criterion variable.

The card pictured below is the model card for the regression model under consideration:

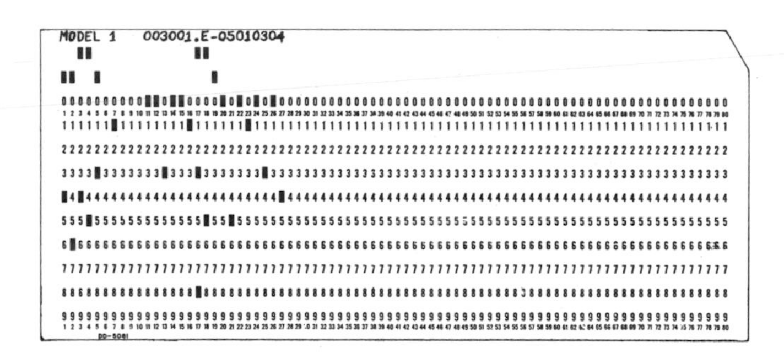

We have arbitrarily identified this model as MODEL 1 in columns 1-10. Let us skip to the column dealing with the stop criterion. For now we will follow our conventional usage of 001.E-05. Starting in column 22, we need to specify the predictors and the criterion. For this model, we are using as predictors the first three variables read off the data card. The criterion is the fourth variable read off the data card. In order to incorporate these variables in the model, we need to have some way to specify them. We do this as illustrated on the model card above. The 010304 indicates that variables one through three will be used as predictors and variable four will be used as the criterion. There are three two-digit fields here; so we must punch 003 in columns 11-13.

> If we should want to construct a model card which represents a model employing predictor variables one and three, with variable five as the criterion, columns 22 and beyond would look like this: 0101030305. This example contains 5 two-digit fields, so 005 should be punched in columns 11-13.

(There will always be an odd number of two-digit fields to the right of the stop criterion because the predictors must be read two two-digit fields at a time followed by one criterion variable, which causes the number in columns 11-13 to be odd.)

The R^2 computed for this model indicates the per cent of variance of the physics exam (variable 4) accounted for by these predictor variables (variables 1, 2, and 3). A more thorough discussion of R^2 was previously presented in Chapter IV.

A restricted model is needed which reflects that the three groups obtained the same score on the physics exam. This model would be obtained by making the restriction that $a_1 = a_2 = a_3$. Since the weights for the three vectors X_1, X_2, and X_3 would be the same, the vectors could be added together to form a new vector. Each element of this vector would be equal to 1, and thus we have a unit vector.

The restricted model needed to answer the research question is:

$$Y_1 = a_0U + E_2$$

This model was shown in Chapter IV to yield an R^2 of zero; and because of this fact, we will not have to punch a card for the restricted model. We will discuss this later.

F Ratio Card (Card #12)

To see if the per cent of variance accounted for by the full model is statistically different than that accounted for by the restricted model, a univariate F test must be run. This program will compute the F ratio if an F ratio card is prepared. The organization of the F ratio card is as follows:

Columns	Information Punched
1-10	Identification (e.g., F ratio 1)
13	6
22-23	The ordinal position of the full model with respect to other model cards.
24-25	The ordinal position of the restricted model with respect to other model cards.
26-29	(df_1) Degrees of freedom in the numerator: ($m_1 - m_2$) where m_1 = the number of linearly independent vectors in the full model and m_2 = the number of linearly independent vectors in the restricted model.
30-33	(df_2) Degrees of freedom in the denominator ($N - m_1$) where N = number of subjects in the study.

Following is the formula which the program uses to compute the F ratio:

$$F = \frac{(R^2{}_f - R^2{}_r)/(m_1 - m_2)}{(1 - R^2{}_f)/(N - m_1)}$$

A discussion of how the degrees of freedom are calculated will be explained more fully later in this chapter.

The computer differentiates a model card from an F ratio card by the number in column 13 on these cards. If this number is odd, the card is read as a model card. (We noted previously that the number of two-digit fields starting in column 22 in a model card will always be odd, and this number appears in column 13.) If this number is even, the card is read as an F ratio card. (We normally use a "6.")

It should be pointed out that the identification information in the first ten columns of the model and F ratio cards is for the experimenter's use only. When the computer has cause to refer to a model by number (as in columns 22-25 of the F ratio card), it will select the model that occurs in that numerical order, whether it is numbered that way in columns 1-10 or not.

The F ratio card for the physics exam problem could look like this:

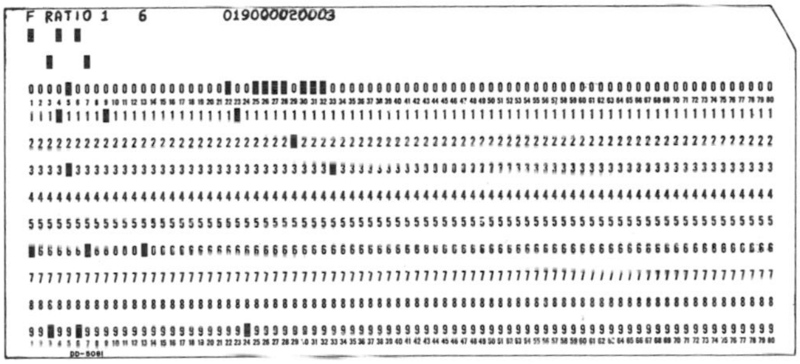

F RATIO 1   6   019000020003

Following the identification (F Ratio 1) and the 6 in column 13, the ordinal sequence number of the full model is specified in columns 22 and 23 as "01." Then, to obtain an $R^2 = 0$, a fictitious model (90) is placed in columns 24-25 to be compared with the full model. Since the program initially zeros all the R^2 values for all models, and we have not computed 90 models, an R^2 value of zero is the value associated with "Model 90."

Next the degrees of freedom in the numerator are specified (0002). The full model, $Y_1 = a_1X_1 + a_2X_2 + a_3X_3 + E_1$, has three unknown weights (a_1, a_2, and a_3) to be calculated. The fictitious model has one weight to be calculated -- the grand mean for all subjects. This is a very difficult idea to understand at first; so do not worry if you have not grasped it yet. Calculating degrees of freedom will be more fully explained in the following problems throughout the book. Therefore, the degrees of freedom in the numerator equal $(m_1 - m_2) = (3 - 1) = 2$, which is punched in columns 26-29 as "0002." The degrees of freedom in the denominator equal $(N - m_1) = (6 - 3) = 3$, which is punched in columns 30-33.

Deck Arrangement

The model card(s) is placed in the program immediately after the last data card and the F ratio card(s) follows the model card. The deck is completed by putting a blank card after the last F ratio card.

The cards relevant to this problem would be organizes as illustrated in Figure 5-3. These cards will follow the last card in Figure 5-1.

```
PHYSICS EXAM N=6  EXAMPLE 1
000060000400004
(2X,3F1.0,X,F2.0)
1 100 04
2 100 06
3 010 10
4 010 12
5 001 20
6 001 20
MODEL 1    003001.E-05010304
F RATIO 1   6         019000020003
```

Figure 5-3. Listing of cards relevant to physics exam problem

These cards, with the program deck preceding them and a blank card following them, can be submitted to the computer.

As a review of all the cards we have discussed, the manner in which information is placed on them, and the order in which each is placed, refer to Figure 5-4. (See pages 90 and 91.)

Output

In order for this method to be of value to you, you must learn to interpret the output which you receive from the computer. The output for the physics exam problem is in Figure 5-5.

```
    PHYSICS EXAM N=6  EXAMPLE 1
MEANS-STANDARD DEVIATIONS-CORRELATIONS
  MEANS
              1          2          3          4
     1     0.3333     0.3333     0.3333    12.0000
  STANDARD DEVIATIONS
              1          2          3          4
     1     0.4714     0.4714     0.4714     6.2183
  CORRELATIONS
              1          2          3          4
     1     1.0000    -0.5000    -0.5000    -0.7960
     2    -0.5000     1.0000    -0.5000    -0.1137
     3    -0.5000    -0.5000     1.0000     0.9097
     4    -0.7960    -0.1137     0.9097     1.0000
```

Figure 5-5. Partial computer output from physics exam example, Problem 5-1

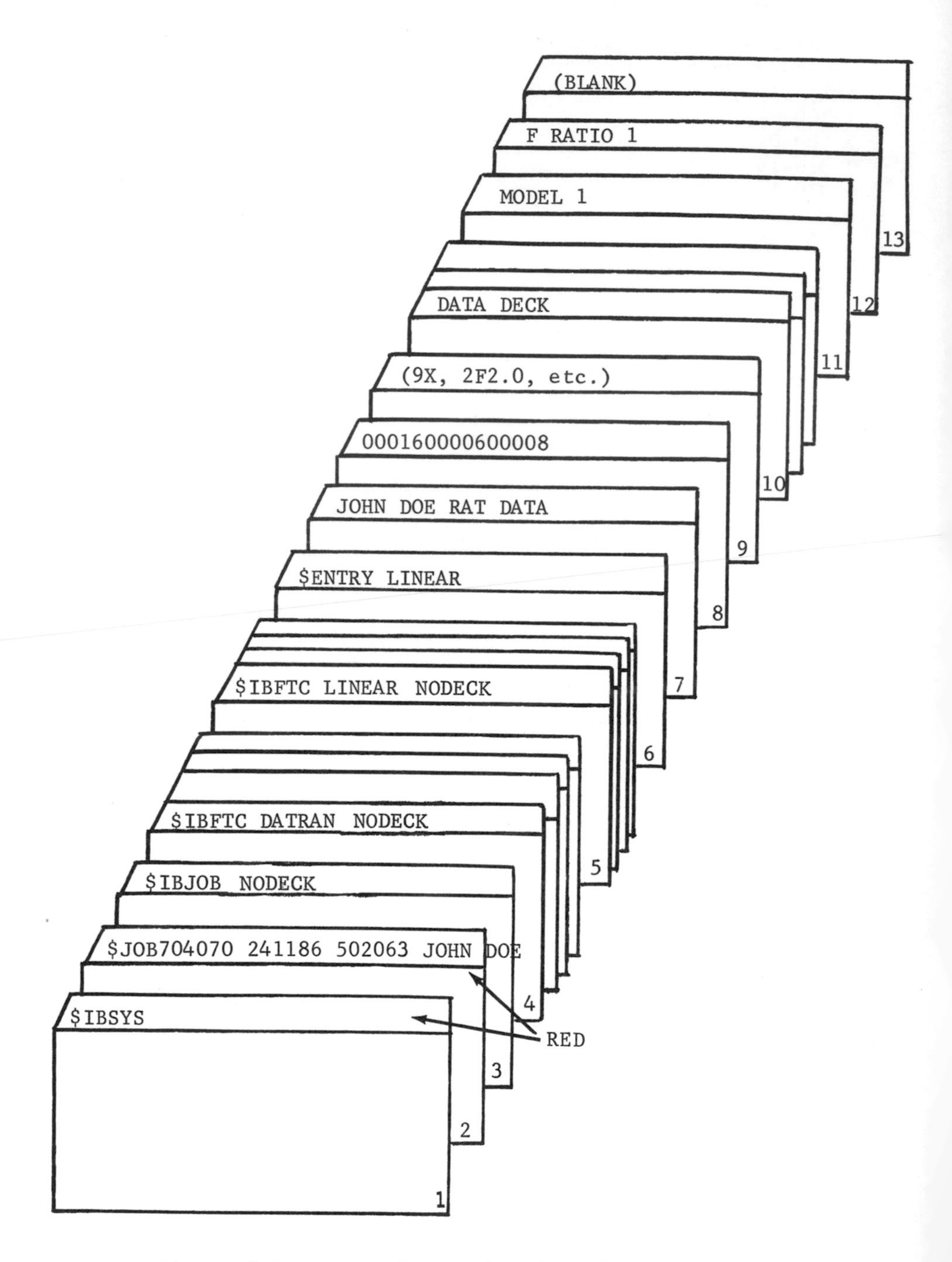

Figure 5-4. General organization of program LINEAR

Explanation of Cards in Figure 5-4

Card 2

Column
19-22 = 7040
28-29 = 70
31-36 = 241186 (Dept. #)
38-42 = 50206 (Prob. #)
44 = 3
46--- = User's name

Card 7 (Title Card)

For identification purposes. Must have a punch somewhere in first twelve columns.

Card 8 (Parameter Card)

Column
1-5 = number of subjects
6-10 = number of variables
11-15 = number of variables plus number of generated variables

Card 9 (Format Card)

Indicates location of data. (aFb.c,X,aFb.c,etc.)

Card 11 (Model Card)

Column
1-10 = identification (e.g., Model 1)
11-13 = number of two-digit fields starting in column 22
14-21 = stop criterion (usually 001.E-05)
22-23, 24-25, 26-27, etc. = specification of predictor variables and criterion variable.

Card 12 (F Ratio Card)

Column
1-10 = identification (F Ratio 1)
13 = 6
22-23 = ordinal sequence for full model
24-25 = ordinal sequence for restricted model
26-29 = df_1 numerator ($m_1 - m_2$) where: m_1 = number of linearly independent vectors in the full model m_2 = number of linearly independent vectors in the restricted model
30-33 = df_2 denominator ($N - m_1$) where N = number of subjects

First, the information on the title card is printed. Next the means, standard deviations, and correlations for each of the variables (vectors) are printed. If this information is all that you need concerning your particular data, a blank card is placed after the last data card and no model or F ratio card is needed. If you have mutually exclusive categories (such as membership in Groups A, B, or C), the sum of the means for those vectors must equal 1. In this case, the sum of the means for vectors 1, 2, and 3 is .9999 (equal to 1, within rounding error), since two subjects or one-third of the total N of 6 is in each group. The mean of the criterion variable (number 4), score on the physics exam is 12.0, which has been shown previously to be the grand mean of the six scores. The standard deviations of mutually exclusive categories are of no use to us; but for continuous vectors, such as variable 4, it gives an indication of the variation of all the scores around the mean. In other words, if the scores are relatively normally distributed, approximately two-thirds of the scores fall within $\pm$ 1 standard deviation of the mean. In this case approximately two-thirds of the scores would be between 5.78 and 18.22 (12.00 $\pm$ 6.22).

The correlation matrix is printed after the standard deviations. The Pearson Product Moment correlation coefficient of each variable with every other variable is calculated and outputted.

If there is a specific question or hypothesis to be tested, model and F ratio cards are added to the deck as explained earlier;

and the information about them will be printed on the output, as shown in Figure 5-6.

```
.....MODEL 1      0.000010
 CRITERION     4
PREDICTORS       1  -    3

   RSQ = 0.98275860                             2

VAR. NUMBER  STD. WT.    ERROR

     1     0.00000000     0.00000001
     2     0.45485881    -0.00000001
     3     1.13714705    -0.00000001

VAR. NUMBER      WEIGHT

     1          0.00000000
     2          5.99999976
     3         14.99999964

CONSTANT=       5.00000036

****F-RATIO = 85.4999  D.F.NUM.= 2. D.F.DEN.=3. 1 0.9828 90 0.0000

                                                          PROB=0.00226

TIME FOR THIS RUN IS      3.40 SECONDS

END-OF-DATA ENCOUNTERED ON SYSTEM INPUT FILE.
```

Figure 5-6. Remaining computer output from physics exam example.

First the identification information (e.g., Model 1) is printed, followed by the stop criterion, which in this case is 10^{-5}, or 0.000010. Below that is the number of the variable which was treated as the criterion variable (4). Then the variables (vectors) which acted as predictors in this model are specified (1-3). Next the RSQ (R^2) is printed: RSQ = .98275860. RSQ ranges from 0.0 to 1.0 and indicates the per cent of variance in the criterion which is accounted for by the predictor variables. In this case 98.27% of the variance in the physics exam is accounted for by the group membership vectors. The standard weights and truncating error made by the computer is indicated, but we won't need to deal with these until later in the text. The weights which are more useful to us are printed after the standard, or beta, weights.

Do you recall the form of the full model?

$$Y_1 = a_0U + a_1X_1 + a_2X_2 + a_3X_3 + E_1$$

The weight for variable 1 (a_1) is equal to 0.000. . . . For variable 2 (a_2) is equal to 5.99 . . . and for variable 3 (a_3) is equal to 14.99. The constant value of 5.00 . . . is the value of a_0 which is multiplied by the unit vector U to give an initial value from which the values for the groups are added or subtracted. The unit vector is automatically included in each model, and as a result, this initial value (a_0) is always found.

You may recall that the means for these groups were calculated in Chapter IV and that they were equal to 5, 11, and 20. These means can be found from these weights if we add the value for the unit vector (Constant = 5.00) to the weight computed for each group variable. So the means are found to be:

Mean of Group 1 = $a_0U + a_1X_1$ = (5.0*1) + (0.0*1) = 5

Mean of Group 2 = $a_0U + a_2X_2$ = (5.0*1) + (5.999*1) = 10.999

Mean of Group 3 = $a_0U + a_3X_3$ = (5.0*1) + (14.999*1) = 19.999

which are within rounding error of the actual group means: 5, 11, 20.

The computer systematically varies the weights for the predictor variables around the constant value to find the largest percentage of variance of the criterion measure for which this combination of predictors can account. In the physics exam example the weights for a_1, a_2, and a_3 are varied by the LINEAR program until the increase in the RSQ is less than the stop criterion (which in this case is 0.00001). The number of times the weights are changed (iterations) is indicated by the number far to the right of the RSQ value. In this case the computer made two iterations, or changed the weights for the predictor variables twice.

Following the information concerning the models, the F ratio calculation is printed. The value of the F ratio is shown as F ratio = 85.4999. Then the degrees of freedom in the numerator and in the denominator which you punched on the F ratio card are printed. In this case the values are 2 and 3, respectively. You may want to check these values with those calculated for this example in Chapter IV. Next the model number of the full model (1) is indicated, followed by the RSQ for that model (0.9828). Then the model number of the restricted model (90) is shown, followed by the RSQ for that model (0.0000). Finally, the probability of obtaining an F ratio of the given magnitude (or larger) with these particular degrees of freedom purely by mathematical chance is indicated. In this case, the probability of finding an F ratio at least as large as 85.4999 with 2 and 3 degrees of freedom is equal to 0.00226. Let us say that we had previously chosen alpha equal to .05 as the level of significance which we would accept. This F ratio would then indicate a significant difference in the means of the three treatment groups, A, B, and C. If all other influencing variables are assumed to be controlled or randomized, as to their effect on the physics exam, we would conclude that the different programed texts used by the three groups caused the differences in the scores on the physics exam.

The next information printed on the output sheet is the amount of time taken by the computer to run the program: 3.40 seconds. Then a statement that the end of the data was encountered on the system input file indicates that everything submitted has been run and no error has been encountered.

We have seen how to prepare the cards for the analysis of the physics exam problem and how to interpret the output from the computer for this problem. Since more practice is undoubtedly necessary, another example will be discussed on the following pages.

This problem is the three-group case presented in the problems in Chapter IV, which dealt with the influence of treatment conditions on test performance.

In this example the treatments were defined as groups receiving one of the following instructions: (1) 10% of the final grade would be determined by the test score; (2) test score would not be counted in the final grade; and (3) 90% of the final grade would be determined by the test score. After we have defined the experimental groups and have collected the criterion measure on each subject in each group, we are ready to punch the data on the data cards. Let us now describe the placement of the data on the card.

We need three columns to identify group membership and another field in which to place the criterion variable. We will assign each subject an identification number which will be a godsend if we, for any reason, happen to lose some data cards, or if we should want to add some additional information to the data cards. It is just a good data processing procedure to assign subject identification numbers, even though they will not be used in any analyses. The position of this information on the data card is actually immaterial because of the flexibility of the LINEAR program, but we will usually place the subject code in the first few columns of the data card. In this example we will choose to use the following columns to represent our data:

Columns	Information Punched
1-2	Subject code.
3-4	Criterion score.
5	1 if a member of Group A; 0 otherwise.
6	1 if a member of Group B; 0 otherwise.
7	1 if a member of Group C; 0 otherwise.

The program cards which we will have to punch for this particular analysis are as follows:

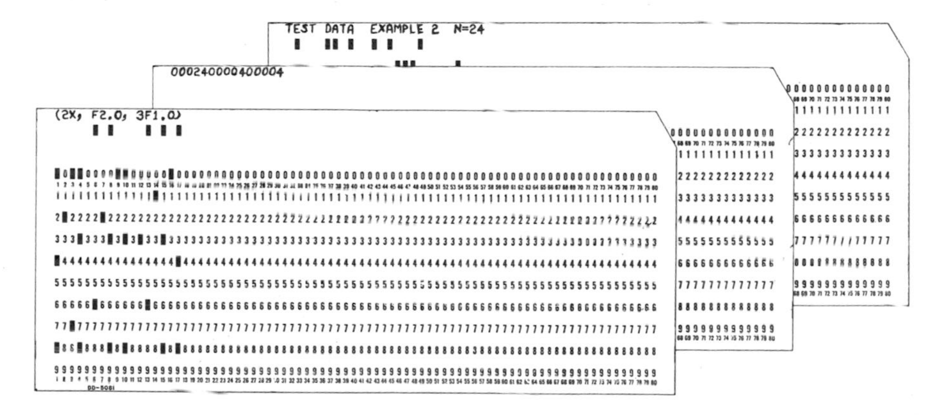

The title card may have any identifying information on it that the researcher so desires. We choose to place the above particular information on the title card for this example.

The parameter card has to have specific information on it. Let us recall that columns 1-5 must contain the number of subjects in this particular problem (24). Columns 6-10 contain the number of variables to be read from the data cards (4). We will again defer discussion of columns 11-15, except to say that columns 11-15 should contain 00004 for this problem (because the total number of variables for this problem is four).

The format card informs the program of how the data is punched on the data cards. Because we do not need the subject code for our analysis, we can skip over this information. We do need to read in the three group membership variables and the criterion variable. Thus we want to skip the first two columns (2X), then read the next two columns because they contain the criterion (F2.0), and finally read the next three columns as single-digit scores (3F1.0).

The remaining cards that we have to punch are the model and F ratio cards. The particular models which would answer the research question were left up to the reader as an exercise at the end of the last chapter. We now present the full and restricted models so as to assist us in constructing our program cards. The research question was: "Do the three experimental groups come from populations which have equal means, i.e., do the three experimental groups yield significantly different mean scores on the criterion variable?"

The full linear model which includes knowledge of group membership is:

$$Y_1 = a_1X_1 + a_2X_2 + a_3X_3 + E_1$$

(Full model for Example II)

where:

Y_1 = the criterion variable

X_1 = 1 if the subject is a member of Group 1; 0 otherwise.

X_2 = 1 if the subject is a member of Group 2; 0 otherwise.

X_3 = 1 if the subject is a member of Group 3; 0 otherwise.

E_1 = the error vector.

a_1, a_2, and a_3 are unknown least squares weights.

The research question tells us to construct the restricted model by assuming that all of the subjects are, in fact, members of one and the same group. Another way of stating this restriction is to indicate that, in fact, we do not have knowledge of group membership, or that we do not know what particular group any one individual belongs to. We know that the three group membership vectors above are mutually exclusive vectors, and we have already learned at the end of Chapter III that the addition of mutually exclusive vectors will yield the unit vector. Thus, our restricted model would be:

$$Y_1 = a_0U + E_2$$

(Restricted model for Example II)

where:

Y_1 = the criterion variable

U = the unit vector

E_2 = the error vector

a_0 = an unknown least squares weight

The reader should note that in the above restricted model, we have no vector which allows us to detect which group a subject belongs to. Thus, we are reflecting in this restricted model the restriction implied by our research question.

The reader should also note that this model will yield an R^2 of zero. That is, by utilizing only the unit vector, we obtain an R^2 of zero. We found this to be the case in the several examples of Chapter IV.

After we have specified our full and restricted models, we are then ready to construct model cards and F ratio cards.

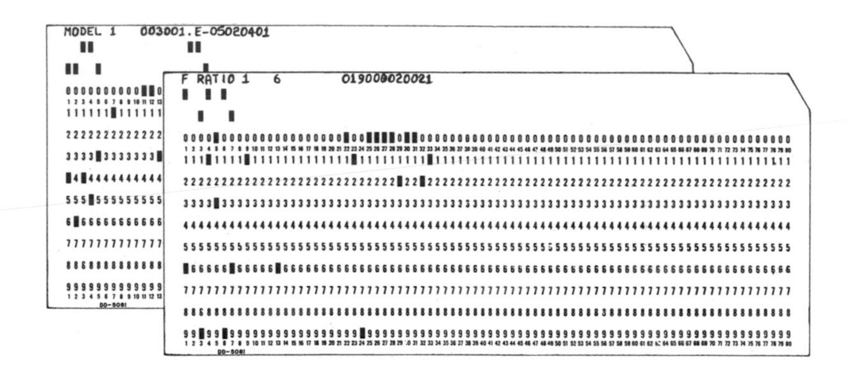

Referring back to Figure 5-4, we construct the above cards in the following manner. The model card is identified as MODEL 1 in columns 1-10. The stop criterion (001.E-05) is placed in columns 14-21. To determine the manner in which we should specify our predictor and criterion variables in columns 22 and following, we refer to our data cards and the manner in which we have stated they will be read (by the format card). In this case, our predictor variables were read as the second, third, and fourth variables. We express this as "0204" (starting in column 22). The criterion variable was read as the first variable, which we express as "01." Since this means we will have three two-digit fields after the stop criterion, we place 003 in columns 11-13; and our model card is complete.

We do not need to construct a model card for the restricted model because we already know that the restricted model will, in fact, yield an R^2 of zero, and we also know that model 90 has been equated with the numerical value of zero. We take this information into account when we need to refer to the restricted model in the F ratio card; so we place "90" in columns 24-25 of the F ratio card.

The F ratio card is identified as F Ratio 1 in columns 1-10. The "6" in column 13 indicates to the computer that this is an F ratio card. We wish to compare the R^2 values of Model 1 and Model 90; so we place "0190" in columns 22-25. The degrees of freedom in the numerator (df_1) is equal to $m_1 - m_2$ (number of linearly independent vectors in the full model minus the number of linearly independent vectors in the restricted). This value is 2 (3-1), and thus we place "0002" in columns 26-29. The degrees of freedom in the denominator (df_2) is equal to $N - m_1$ (or in this case 24 - 3). We place "0021" in columns 30-33 and the F ratio card is complete.

The cards can now be assembled as shown in Figure 5-4 and submitted to the computer. The output of this particular problem should be as shown in Figure 5-7.

```
  TEST DATA  EXAMPLE 2  N = 24

MEANS-STANDARD DEVIATIONS-CORRELATIONS

  MEANS
               1          2          3          4
    1     5.7083     0.3333     0.3333     0.3333
STANDARD DEVIATIONS
               1          2          3          4
    1     2.3888     0.4714     0.4714     0.4714
CORRELATIONS
               1          2          3          4
    1     1.0000    -0.2837    -0.3207     0.6043
    2    -0.2837     1.0000    -0.5000    -0.5000
    3    -0.3207    -0.5000     1.0000    -0.5000
    4     0.6043    -0.5000    -0.5000     1.0000

.....MODEL 1      0.000010
 CRITERION    1
PREDICTORS      2 -  4

   RSQ = 0.36568299

VAR. NUMBER   STD. WT.    ERROR

    2        0.02466693      0.00000000
    3        0.00000000      0.00000000
    4        0.61667331     -0.00000001

VAR. NUMBER     WEIGHT

    2        0.12499998
    3        0.00000000
    4        3.12499997

CONSTANT=     4.62500000

****F-RATIO=6.0532  D.F.NUM.=2. D.F.DEN.=21. 1 0.3657 90 0.0000 PROB=.00840

TIME FOR THIS RUN IS      7.02 SECONDS

END-OF-DATA ENCOUNTERED ON SYSTEM INPUT FILE
```

Figure 5-7. Computer output for Example II.

The first information printed is the title card. Again this information helps to identify the particular analysis. Then comes information which will always appear on LINEAR output: MEANS --

STANDARD DEVIATIONS -- CORRELATIONS. Next we find the means of our four variables. We should always check this information against what we would expect to obtain. We have some idea as to the value of the criterion mean; we know that it will be less than 20, for instance. We can also check the means for the three group vectors, for these means will actually be the number of people in the group divided by the total number of people (symbolically: group N/total N). In this particular example we have eight subjects in each group, and therefore, the means of variables 2, 3, and 4 should all be 0.3333 (8/24).

The correlation matrix is also printed automatically and contains the intercorrelations of all of the variables. Often this is valuable and needed information, but for the purposes of this problem, we do not need to deal with the correlation matrix.

We next find the results of our generated models. The models are first identified by the information which we punched in columns 1-10. The criterion variable is then identified and the predictor variables are specified on the next line of output.

RSQ is an abbreviation for R^2, or the per cent of variance in the specified criterion that is accounted for by the specified predictor variables. We will ignore the remaining information and focus now on the F ratio line of information. First, the calculated F ratio is printed, with its associated degrees of freedom for the numerator (D.F. NUM. = 2) and degrees of freedom for the denominator (D.F. DEN. = 21). The sequence number of the full model being tested (1) is printed along with the actual magnitude of the R^2

associated with this model (.3657). The same information is then presented for the restricted model (sequence number of 90 and an R^2 of 0.0000).

We are then given the probability (PROB=.00840) of an F ratio this large or larger occurring by chance alone. If the probability value is less than our pre-set alpha level (which we have again selected as .05 in this case), then we can reject the hypothesis of no difference between treatments and accept the hypothesis that there is a difference between treatment means. Since the outputted probability level is less than .05, we can accept the hypothesis that the nature of the instructions have a differential effect upon test performance.

We are also given the information as to how much time it took the computer to do all the calculations it was told to do. After using the LINEAR program at your own installation a few times, you will become familiar with the amount of time that it should take for problems of various sizes.

Let us return to the WEIGHT section of the output to see if we can utilize any more of the information given to us. The weights given for the three variables (group membership vectors) are the "unknown least squares weights" that we have so often referred to. We have previously alluded that these weights should reflect the group means, and these certainly do not, for the mean of Group 2 was not 0.0, but $[(4 + 4 + 3 + 8 + 7 + 4 + 2 + 5)/8]$ or 4.625. This value of 4.625 is the value of the "CONSTANT" on the output, but what is this constant?

For purposes of calculating an F-test using program LINEAR, the unit vector must appear in both regression models. The unit vector does appear in our above restricted model, but does not appear in our full model. But what happens if we simply "add in" a unit vector to the full model for this problem? Let us again write the full model below:

$$Y_1 = a_1X_1 + a_2X_2 + a_3X_3 + E_1$$

(Full model for Example II)

Now adding in the unit vector, this model becomes:

$$Y_1 = b_0U + b_1X_1 + b_2X_2 + b_3X_3 + E_1$$

(Full model for Example II, with unit vector)

Since vectors X_1, X_2, and X_3 are mutually exclusive, and when added together equal the unit vector, the vector U is actually the sum of a linear combination of X_1, X_2, and X_3. Thus, when we add the unit vector to the full model, we do not add in any information that we do not already have in the full model.

Let us look at the expected values (means) for the three groups without the unit vector:

$$\widetilde{Y}_1 = a_1X_1 + a_2X_2 + a_3X_3$$

The expected value for members of Group 1 is:

$$a_1(1) + a_2(0) + a_3(0) = a_1(1) = a_1$$

The expected value for members of Group 2 is:

$$a_1(0) + a_2(1) + a_3(0) = a_2(1) = a_2$$

The expected value for members of Group 3 is:

$$a_1(0) + a_2(0) + a_3(1) = a_3(1) = a_3.$$

$$a_1 \begin{bmatrix} 1\\1\\1\\1\\1\\1\\1\\1\\0\\0\\0\\0\\0\\0\\0\\0\\0\\0\\0\\0\\0\\0\\0\\0 \end{bmatrix} + a_2 \begin{bmatrix} 0\\0\\0\\0\\0\\0\\0\\0\\1\\1\\1\\1\\1\\1\\1\\1\\0\\0\\0\\0\\0\\0\\0\\0 \end{bmatrix} + a_3 \begin{bmatrix} 0\\0\\0\\0\\0\\0\\0\\0\\0\\0\\0\\0\\0\\0\\0\\0\\1\\1\\1\\1\\1\\1\\1\\1 \end{bmatrix}$$

For this particular model, a_1 is, in fact, the mean of Group 1 (4.75), a_2 is the mean of Group 2 (4.625), and a_3 is the mean of Group 3 (7.75).

Now we want to ask: "What are the expected values for the three groups for the full model when we have added in the unit vector?"

$$\widetilde{Y}_1 = b_0U + b_1X_1 + b_2X_2 + b_3X_3$$

The expected value for members of Group 1 is:

$$b_0(1) + b_1(1) + b_2(0) + b_3(0) = b_0+b_1$$

The expected value for members of Group 2 is:

$$b_0(1) + b_1(0) + b_2(1) + b_3(0) = b_0+b_2$$

The expected value for members of Group 3 is:

$$b_0(1) + b_1(0) + b_2(0) + b_3(1) = b_0+b_3$$

$$b_0 \begin{bmatrix} 1 \\ 1 \\ 1 \\ 1 \\ 1 \\ 1 \\ 1 \\ 1 \\ 1 \\ 1 \\ 1 \\ 1 \\ 1 \\ 1 \\ 1 \\ 1 \\ 1 \\ 1 \\ 1 \\ 1 \\ 1 \\ 1 \\ 1 \\ 1 \end{bmatrix} + b_1 \begin{bmatrix} 1 \\ 1 \\ 1 \\ 1 \\ 1 \\ 1 \\ 1 \\ 1 \\ 0 \\ 0 \\ 0 \\ 0 \\ 0 \\ 0 \\ 0 \\ 0 \\ 0 \\ 0 \\ 0 \\ 0 \\ 0 \\ 0 \\ 0 \\ 0 \end{bmatrix} + b_2 \begin{bmatrix} 0 \\ 0 \\ 0 \\ 0 \\ 0 \\ 0 \\ 0 \\ 0 \\ 1 \\ 1 \\ 1 \\ 1 \\ 1 \\ 1 \\ 1 \\ 1 \\ 0 \\ 0 \\ 0 \\ 0 \\ 0 \\ 0 \\ 0 \\ 0 \end{bmatrix} + b_3 \begin{bmatrix} 0 \\ 0 \\ 0 \\ 0 \\ 0 \\ 0 \\ 0 \\ 0 \\ 0 \\ 0 \\ 0 \\ 0 \\ 0 \\ 0 \\ 0 \\ 0 \\ 1 \\ 1 \\ 1 \\ 1 \\ 1 \\ 1 \\ 1 \\ 1 \end{bmatrix}$$

Now let us check these expectations with the computer output, realizing that b_0 is the "constant" and is equal to 4.625; b_1 = 0.12499; b_2 = 0.000; b_3 = 3.12499. Therefore, the expected value of Group 1 is: $b_0 + b_1 = 4.625 + 0.12499 = 4.7499 \stackrel{a}{=} 4.75$. The expected value of Group 2 is: $b_0 + b_2 = 4.625 + 0.000 = 4.625$. Finally, the expected value of Group 3 is: $b_0 + b_3 = 4.625 + 3.12499 = 7.7499 \stackrel{a}{=} 7.75$. These expectations are only approximate (but nevertheless, very close) because of the iteration procedure used by the LINEAR program.

Thus, we have shown that the addition of the unit vector to the full model does not, in fact, change the expectations, nor does it change the value of the R^2 (note that the error vector is the same in both models). The addition of the unit vector does allow us to compute the F ratio via the R^2 values. We have also shown that the group means can still be calculated when the unit vector is inserted into a regression model. We simply have to add this "constant" to the least squares weight of each group vector.

The reader should be made aware of the fact that the unit vector is added automatically by the LINEAR program to every model that is entered on a model card. The researcher thus has to take this into consideration when computing his degrees of freedom. In this particular example, the unit vector does not affect the previously calculated degrees of freedom, for it is a linear combination of the three mutually exclusive group membership vectors. Thus, when including the unit vector, we still have three linearly independent vectors in the full model. The unit vector is also a linear combination of the one vector in the restricted model; in fact, it is the same vector, for the sole vector in the restricted model was the unit vector.

> Note that when you count the number of linearly independent vectors for a model, the unit vector must also be considered. When mutually exclusive vectors are used as predictor variables, however, you do not count the unit vector, since the unit vector can be shown to be linearly dependent upon the mutually exclusive vectors in the set of predictor variables.

Now that you have read a detailed description of how to prepare two problems for computer analysis, try your hand at the following exercises.

Problems for Chapter V

1. Punch all the cards (including data cards) necessary to run the LINEAR program for data for the physics exam example and submit it to the computer.

2. Do the same thing for the test conditions example (data on pages 71 and 72).

3. Write the title and parameter cards for:

   (a) 36 subjects with cards read by the format (3X, 5F1.0, 3F2.0, 2X, 2F1.0, 2F3.0).

   (b) 180 subjects with cards read by the format (5X, 2F1.0, 15F2.0, 3F1.0).

   (c) 1,000 subjects with cards read by the format (2X, F2.0, 3X, F2.0, 3F1.0, 2F3.0).

4. Referring to 3(a) above, write model cards reflecting the following:

   (a) variables 1, 3, 5, 6, 7, and 8 as predictors and variable 2 as criterion.

   (b) variables 8, 9, 11, and 12 as predictors and variable 5 as criterion.

   (c) variables 2, 4, 5, and 11 as predictors and variable 7 as criterion.

   (d) variables 2, 4, and 5 as predictors and variable 7 as criterion.

   (e) variable 2 as predictor and variable 7 as criterion.

5. Referring to 4 above and assuming that we entered those models in that order, write F ratio cards which will reflect the following: (You may assume that all the predictors are mutually exclusive vectors.)

(a) 4(c) as full model and 4(d) as restricted model.

(b) 4(c) as full model and 4(e) as restricted model.

(c) 4(d) as full model and 4(e) as restricted model.

6. Write title, parameter, format, model, and F ratio cards for the following: We have data cards for 50 subjects which are punched as follows:

Columns	
1-2	Subject number.
3	1 if male; 0 otherwise.
4	1 if female; 0 otherwise.
5-6	Pre-score on reaction time test.
7-8	Post-score on reaction time test.

We want to know if there is a significant difference between the performance of males and that of females as measured by the pre-test.

As a second research question, we want to know if there is a significant difference of the same sort as measured by the post-test. Note that one set of title, parameter, and format cards will suffice for these two questions; but you will need to write additional model and F ratio cards for the second.

Chapter VI Complex Models for Regression Analysis

Up to this point we have been concerned with problems in which the predictor variables were mutually exclusive, and those problems conformed to conventional analysis of variance. This chapter introduces problems which use either continuous predictors or a combination of mutually exclusive predictors and continuous predictors.

Problems involving curvilinear relationships are presented and linear equations are developed to reflect the non-rectilinear functions. Second, third, and higher order polynomials are shown to be possible ways to reflect the state which one might expect from behavioral theory. In conjunction with the use of complex models, procedures are presented for use in cross-validation of research findings.

Multivariate use of categorical predictors is discussed, and problems similar to conventional analysis of variance are solved using multiple linear regression analysis.

Two empirical studies are summarized to show the use of the several notions discussed in this chapter.

Linear Equations Which Minimize Errors of Prediction

In order to introduce the use of continuous predictor vectors, let us consider one equation. This simple equation is fundamental to all subsequent discussion:

$$\tilde{Y} = a + bX$$

Most readers will recognize this equation as the definition of a line, where X and $\tilde{Y}$ are positions along the X and Y axes; a is the Y-intercept; and b is the slope of the line. This equation is useful to us in multiple linear regression analysis, re-defining the elements as follows:

Y = the observed criterion score

$\tilde{Y}$ = predicted criterion score

a = regression constant

b = partial regression weight which reflects the amount of increase in Y for each unit increase in X

X = the predictor variable

The constant a and b are calculated such that $\Sigma(Y-\tilde{Y})^2$ is minimized (this is the "ESS" presented in Chapter IV).

The relationship between the two ways of viewing the equation will become clearer as we go along. To illustrate, let us consider a few figures that focus upon the various components in the equation.

Figure 6-1 shows the scores of five students on two tests (X and Y). Student A received a zero score on both tests; Student B received a score of "1" on Test X and also a "1" on Test Y, etc. If we wish to cast a line which best reflects the relation between X and Y, it is obvious that the line will go through the five points (see Figure 6-2).

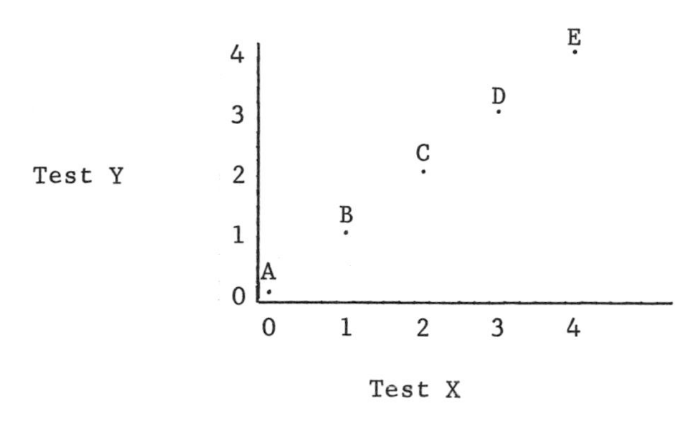

Figure 6-1. Students A through E completed Test X and Test Y. The points reflect the relation of the two scores.

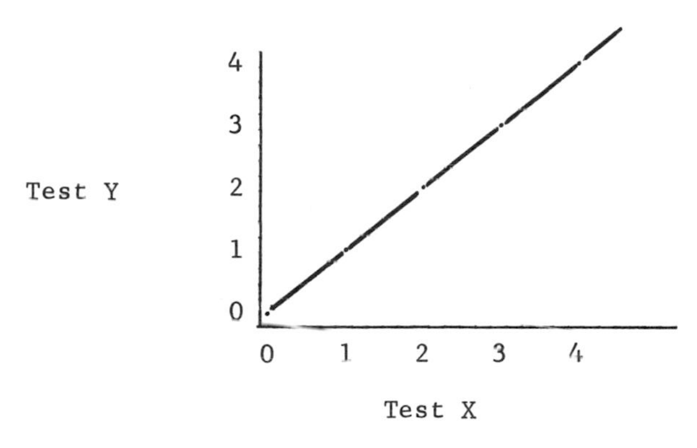

Figure 6-2. The line of best fit between Test X and Test Y for five students.

What values for a and b in the equation $\tilde{Y} = a + bX$ will best reflect this line? First, you might note that in all five cases, each unit increase on Test X is accompanied by a one-unit increase on Test Y. The weight b in the equation essentially reflects the slope of increase in Y as we increase X. In this case there is a one-to-one correspondence between these two changes in scores; therefore, the b weight is 1:

$$\tilde{Y} = a + (1)X.$$

What value is necessary for the weight a, such that $\Sigma(Y-\tilde{Y})^2$ is minimized? When $\Sigma(Y-\tilde{Y})^2$ is minimized, the line then best represents the points being considered. In this case, the units in X and Y are identical and no adjustment is necessary. For example, let us calculate $\tilde{Y}$ for each subject using the following weights, a = 0; b = 1: where $\tilde{Y}$ = predicted score on Y.

$$\tilde{Y} = a + bX$$

$$\tilde{Y}_A = 0 + (1)0 = 0$$

$$\tilde{Y}_B = 0 + (1)1 = 1$$

$$\tilde{Y}_C = 0 + (1)2 = 2$$

$$\tilde{Y}_D = 0 + (1)3 = 3$$

$$\tilde{Y}_E = 0 + (1)4 = 4$$

Student A scored zero on Test X; therefore, the weight 1 (b) times zero equals zero and combined with an "a" weight of zero, Student A's predicted score ($\tilde{Y}_A$) on Y equals zero. The same substitution was made for $\tilde{Y}_B$, etc. In this special case, you will note that $\Sigma(Y-\tilde{Y})^2$ equals zero:

$$Y_A - \tilde{Y}_A = (0 - 0) = 0$$

$$Y_B - \tilde{Y}_B = (1 - 1) = 0$$

$$Y_C - \tilde{Y}_C = (2 - 2) = 0$$

$$Y_D - \tilde{Y}_D = (3 - 3) = 0$$

$$Y_E - \tilde{Y}_E = (4 - 4) = 0$$

The squaring of the difference between $Y-\tilde{Y}$ was not carried out since all differences equaled zero. Let us consider several other figures to illustrate the weights a and b.

In Figure 6-3 four points are shown to reflect the relation of the scores on Tests X and Y for Students F through I. Student F scored "1" on Test X and "4" on Test Y. The four points fall on one line; therefore, you should suspect that when a and b are calculated, no error will be found.

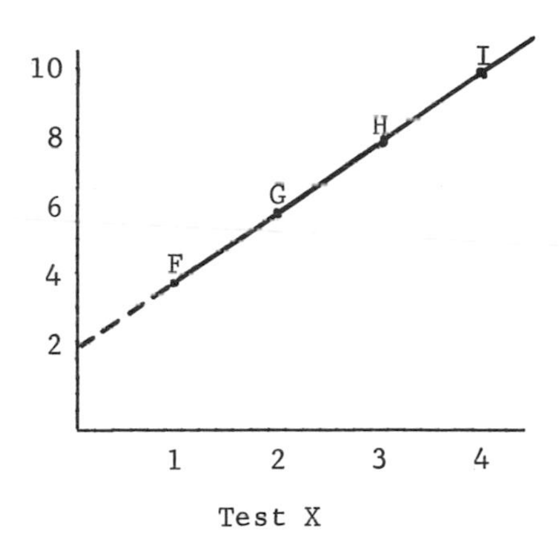

Figure 6-3. A line with four points which indicate the relation of scores on Test X and Test Y for Students F, G, H, and I.

What relationship exists between X and Y? A one-point increase on X (from 1 to 2) is accompanied by a two-point increase on Y (from 4 to 6). In that all points fall on the same line, a one-point increase from 3 to 4 on X is also accompanied by a two-point increase on Y (8 to 10). Since the weight b reflects the amount of increase on Y for each unit increase on X, the weight b equals 2.

If we set a = 0 and b = 2 in the equation $\tilde{Y} = a + bX$, we would obtain the following four predicted scores:

$$\tilde{Y}_F = 0 + 2(1) = 2$$

$$\tilde{Y}_G = 0 + 2(2) = 4$$

$$\tilde{Y}_H = 0 + 2(3) = 6$$

$$\tilde{Y}_I = 0 + 2(4) = 8$$

We can get $\Sigma(Y-\tilde{Y})^2$ by finding the difference between Y_F and $\tilde{Y}_F$, etc.:

	$Y - \tilde{Y}$	$Y - \tilde{Y}$	$(Y-\tilde{Y})^2$	
F =	4 - 2	+2	4	
G =	6 - 4	+2	4	$\Sigma(Y-\tilde{Y})^2 = 16$
H =	8 - 6	+2	4	
I =	10 - 8	+2	4	

By setting a = 0, less than perfect prediction is obtained. If you look at Figure 6-3, an extrapolation of the line is made to intercept the Y axis (dotted line). The point of intercept is 2. Likewise, in the calculations given above, the difference between each observed and predicted score is 2. If we set a = 2, therefore adding a constant to each score, the difference between Y and $\tilde{Y}$ equals zero for each observation and the $\Sigma(Y-\tilde{Y})^2$ is minimized (we cannot get better than perfect prediction). The weights a = 2 and b = 2 are the only two weights which satisfy the condition where $\Sigma(Y-\tilde{Y})^2$ is minimal.

These two examples are, of course, idealized; we seldom find perfect prediction.

The intent of this discussion was to introduce intuitively the notion that b represents the slope of increase of Y across X and that the constant, a, is the value of Y where the line intercepts the Y axis.

Prior to getting down to the business of hypothesis testing, let us entertain one more idealized problem -- this time where prediction is <u>not</u> perfect.

Suppose that we have a situation in which the scores for individuals on Test P and Test Q are plotted as follows:

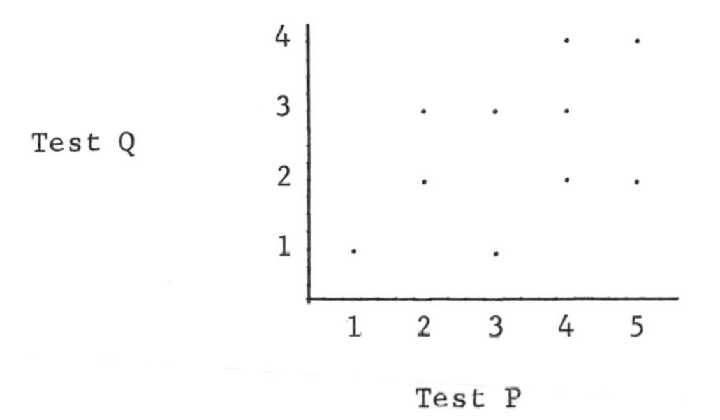

It is obvious that no line will perfectly represent these points. But we can still find a line that best represents it. The line of best fit is defined as that straight line which minimizes the sum of the squared deviations of the points from that line.

Figure 6-4 shows the line of best fit (such that $\Sigma(Y-\widetilde{Y})^2$ is minimized) between scores on Tests X and Y for six students. You will note that Student A scored 2 on Test X and 1 on Test Y. Student B also scored 2 on Test X but scored 3 on Test Y. Student C had scores of 4 and 2, while Student D had scores of 4 and 4. It should be obvious that no straight line can be drawn through points A, B, C, and D because at point 2 on the X axis, A and B are not occupying the same point on Y. The same condition exists for C and D as well as for E and F. We can, however, cast a line among the points to minimize $\Sigma(Y-\widetilde{Y})^2$. Since this is an idealized example, the values for a and b can be intuitively calculated. The symmetry of the distance between each pair allows us to cast a straight line which is the mid-point between each pair of students.

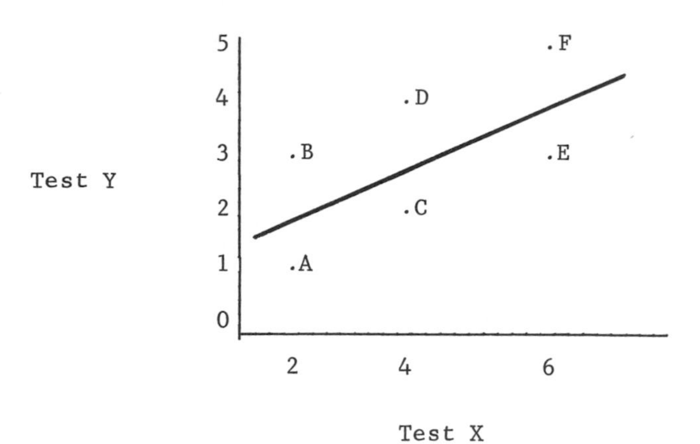

Figure 6-4. The line of best fit between X and Y with six students represented by points around the line.

The line of best fit is presented in Figure 6-4. The weight b can be determined by noting that for a two-unit increase in X (from 2 to 4), the line reflects a one-unit increase in Y (from 2 to 3). Therefore, b should equal 1/2. The extropolated line (dotted) intercepts the Y axis at point 1; therefore, a = 1.

To predict $\tilde{Y}$ such that $\Sigma(Y-\tilde{Y})^2$ is minimal, the equation should be:

$$Y = 1 + 1/2\ X$$

Using these weights to predict each student's scores, we can get:

$$\tilde{Y}_A = 1 + \tfrac{1}{2}(2) = 2$$

$$\tilde{Y}_B = 1 + \tfrac{1}{2}(2) = 2$$

$$\tilde{Y}_C = 1 + \tfrac{1}{2}(4) = 3$$

$$\tilde{Y}_D = 1 + \tfrac{1}{2}(4) = 3$$

$$\tilde{Y}_E = 1 + \tfrac{1}{2}(6) = 4$$

$$\tilde{Y}_F = 1 + \tfrac{1}{2}(6) = 4$$

Furthermore, we can calculate $\Sigma(Y-\tilde{Y})^2$ in the following manner. Can you find a better set of weights? Try a few different weights.

Student	Y	$\tilde{Y}$	$(Y-\tilde{Y})$	$(Y-\tilde{Y})^2$	
A	1	2	-1	1	
B	3	2	+1	1	
C	2	3	-1	1	$\Sigma(Y-\tilde{Y})^2 = 6$
D	4	3	+1	1	
E	3	4	-1	1	
F	5	4	+1	1	

Figure 6-4a.

The LINEAR program presented in Chapter V calculates the constant a and the partial regression weight b so as to minimize $\Sigma(Y-\tilde{Y})^2$. (A formal discussion of regression weight calculation can be found in Guilford (1965), Chapter 15.)

Predicting from Continuous Data: Intuitive Approach

The research position taken in the book is based upon the assumption that we wish to account for variations in human behavior. In Chapter IV, we raised the question: "Can knowledge of programed treatment mode help us to account for performance on tests designed to assess achievement in physics?" The analysis of variance was used to assess the effectiveness of the programs. In Chapter IV the knowledge was cast into mutually exclusive categories. We are now considering another approach that we might use to account for behavior, one that includes the utilization of continuous data. Look at the data summarized in Figure 6-4. Suppose the test data called "X" was a measure of "intelligence" (whatever that may be). We might wish to ask: "Can information of intelligence allow us to predict achievement in an English class (as measured by Y)?" The data presented in Figure 6-4 can be represented in the familiar vector form. The equation $Y = a_0U + a_1X_1 + E_1$ should look familiar;

where:

Y = observed score on the achievement test.

X_1 = a vector in which the elements are the score on the intelligence corresponding to the score on the criterion achievement test.

U = the unit vector.

a_0 = when multiplied by U yields the regression constant.

a_1 = partial regression weight calculated so as to minimize the sum of the squared elements in E_1.

E_1 = a vector whose elements are the difference between the observed score and predicted score $(Y-\tilde{Y})$.

Figure 6-5 shows the vector representation of these data.

$$\overset{Y}{\begin{bmatrix} 1 \\ 3 \\ 2 \\ 4 \\ 3 \\ 5 \end{bmatrix}} = a_0 \overset{U}{\begin{bmatrix} 1 \\ 1 \\ 1 \\ 1 \\ 1 \\ 1 \end{bmatrix}} + a_1 \overset{X_1}{\begin{bmatrix} 2 \\ 2 \\ 4 \\ 4 \\ 6 \\ 6 \end{bmatrix}} + \overset{E_1}{\begin{bmatrix} ? \\ ? \\ ? \\ ? \\ ? \\ ? \end{bmatrix}}$$

Figure 6-5. The vectors for Y, U, X_1, and E_1 related to data of intelligence and achievement.

If we substitute the value for a_0 (which happens to be equivalent to a in the general equation of a line), we find the constant to be one (1). Recall that this is the Y-intercept of the line.

Furthermore, if we substitute b for a_1 (the partial regression weight) derived earlier, we have $a_1 = \frac{1}{2}$. We can now calculate E_1; see Figure 6-6.

$$\underset{Y}{\begin{bmatrix}1\\3\\2\\4\\3\\5\end{bmatrix}} = 1\underset{U}{\begin{bmatrix}1\\1\\1\\1\\1\\1\end{bmatrix}} + .5\underset{X_1}{\begin{bmatrix}2\\2\\4\\4\\6\\6\end{bmatrix}} + \underset{E_1}{\begin{bmatrix}-1\\+1\\-1\\+1\\-1\\+1\end{bmatrix}}$$

Figure 6-6. The values for E_1 calculated with the weights $a_0 = 1$ and $a_1 = .5$.

You will note the elements in E_1 are identical to the elements in vector $(Y-\tilde{Y})$ presented in Figure 6-4a. Indeed, $E_1 = (Y-\tilde{Y})$, since we have changed only the notation. If we square each element in E_1 and sum this, we have the ESS_w for an analysis of variance. It may sound ridiculous to call X_1 (a continuous vector) a group; yet if one can perceive a continuous vector as an infinite number of groups, one for each point on the continuum, then to call the error term "within" seems reasonable. An alternative, which seems easier for the authors, is to call the ESS_w the error sums of squares <u>with knowledge</u>. When we say "with knowledge," we can mean: (1) knowledge of group membership; (2) knowledge of some continuous sort, such as intelligence measure; or (3) any combination of vector information.

Furthermore, when a restriction is placed upon a model, we can call the new error sums of squares ESS "<u>without knowledge</u>" and this can be used as ESS_t.

For example, let us hypothesize that the weight $a_1 = 0$ for the equation:

$$Y = a_0U + a_1X_1 + E_1$$

A vector multiplied by 0 gives a null vector; therefore, the restricted model can be expressed:

$$Y = a_0U + E_2$$

Figure 6-7 shows the vector and the sum of squared elements in E_2.

$$\underset{Y}{\begin{bmatrix}1\\3\\2\\4\\3\\5\end{bmatrix}} = 3\ \underset{U}{\begin{bmatrix}1\\1\\1\\1\\1\\1\end{bmatrix}} + \underset{E_2}{\begin{bmatrix}2\\0\\1\\1\\0\\2\end{bmatrix}} \qquad \underset{E_2^{\,2}}{\begin{bmatrix}4\\0\\1\\1\\0\\4\end{bmatrix}}$$

$$\Sigma E_{2_i}^{\,2} = 10$$

Figure 6-7. The weight for $a_0 = 3$ is the grand mean which minimizes the sum of the elements in E_2 squared.

The F ratio can be calculated by substituting in the formula:

$$F = \frac{(ESS_t - ESS_w)/\ df_1}{ESS_w\ /\ df_2} \qquad N = 6$$

where:

$$ESS_t = \Sigma E_{2_i}^{\,2} = 10$$

$$ESS_w = \Sigma E_{1_i}^{\,2} = 6$$

$$df_1 = 2 - 1 = 1$$

$$df_2 = N - 2 = 4$$

$$F = \frac{(10 - 6)/\ 1}{6\ /\ 4} = \frac{4}{1.5} = 2.67$$

With one and four degrees of freedom, an F of 2.67 is likely to occur quite often. An F value of 7.71 or larger is likely to be encountered five times in 100. Theoretically, we can conclude that although some linear relationships can be visually presented, the relationship can easily be attributed to sampling error. Essentially, the overlap in scores does not allow for much of a reduction of total error variation (i.e., there is a subject with an X value of 4 who has a Y value less than a subject who has an X value of 2).

Predicting from Continuous Data: Formal Approach

We now illustrate how regression weights can be computed from the raw data when the predictor variable is a continuous vector, rather than a dichotomous vector. We will focus on the simple case of only one continuous predictor in conjunction with the unit vector. What we are actually attempting to do is to find the line of best fit in conjunction with the unit vector. The line of best fit is defined as that straight line which minimizes the sum of the squared deviations of the points from the line.

Let's look at Figure 6-8.

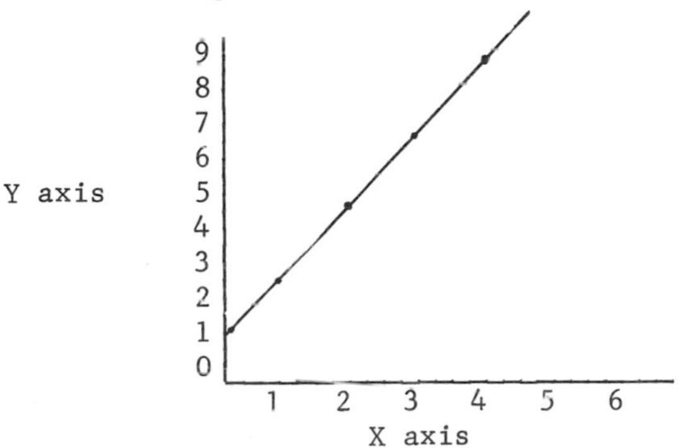

Figure 6-8. Bivariate plot with perfect prediction.

Almost everyone would agree on the line of best fit for Figure 6-8. We have perfect prediction and thus the sum of squared deviations will be equal to zero. We can predict each value of Y from the known values of X via the following formula: $Y = 1 + 2X$. This equation is indeed a linear equation because it fits the format of the general equation for a straight line: $Y = a + bX$. We note that the value of a is actually the Y-intercept, or the value on the Y axis where the line crosses the Y axis. "b" also has some meaningful interpretation and is called "the slope of the line," which represents the amount of change in Y for every unit change in X. Thus, in this particular example, when we move one unit along the X axis, we move two ($b = 2$) units along the Y axis.

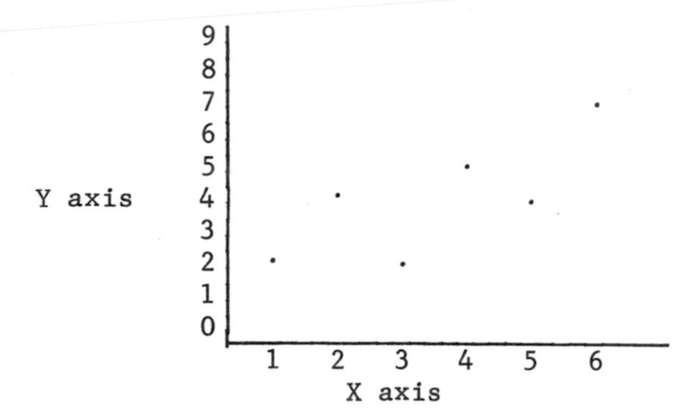

Figure 6-9. Bivariate plot without perfect prediction.

Now look at Figure 6-9. The placement of the regression line is now not so apparent. Two people may not agree on the line of best fit, and that is why we have to resort to computing the line of best fit from the data itself.

Let's take the general equation for a straight line and investigate the process.

1. $Y = a + bX$ (General equation for a straight line). Now we can represent a predicted Y score by $\tilde{Y}$, and thus we have:
2. $\tilde{Y} = a + bX$ (General equation for predicting a score). Now remember that our criterion for constructing this line of best fit was to minimize the sum of the squared deviations of the predicted score from the actual observed criterion score, or:
3. $\Sigma(Y-\tilde{Y})^2$ at a minimum (Expression which is to be minimized in method of least squares). Substituting (2) into (3), we get:
4. $\Sigma[Y - (a+bX)]^2$. (Expression which is to be minimized in method of least squares). We now have to call upon calculus to solve this problem for us; so we will simply give the final equations for a and b.

$$(6.4) \quad b = \frac{\Sigma XY - \frac{(\Sigma X)(\Sigma Y)}{N}}{\Sigma X^2 - \frac{(\Sigma X)^2}{N}}$$

(Regression weight for continuous vector - "slope")

$$(6.5) \quad a = \bar{Y} - b\bar{X}$$

(Regression weight for unit vector - "Y-intercept")

X	Y	X^2	XY
1	2	1	2
2	4	4	8
3	2	9	6
4	5	16	20
5	4	25	20
6	7	36	42
$\Sigma X = 21$	$\Sigma Y = 24$	$\Sigma X^2 = 91$	$\Sigma XY = 98$

$(\Sigma X)^2 = 441$ $\quad \bar{Y} = 24/6 = 4$

$\bar{X} = 21/6 = 3.5$

$$b = \frac{\Sigma XY - \frac{(\Sigma X)(\Sigma Y)}{N}}{\Sigma X^2 - \frac{(\Sigma X)^2}{N}} = \frac{98 - \frac{(21)(24)}{6}}{91 - 441/6} = \frac{98-84}{91-73.5} = \frac{14}{17.5} = 8$$

$$a = \bar{Y} - b\bar{X} = 4-(.8)(3.5) = 4-2.8 = 1.2$$

Figure 6-10

Now we show how these coefficients can be solved for a particular set of data. Figure 6-10 contains the data from Figure 6-9. The calculations for a and b are shown in Figure 6-10. We can run a quick check on the computed a and b values, because we know that the mean of the X values ($\overline{X}$) should predict the mean of the Y values ($\overline{Y}$). Thus, if we place the value of $\overline{X}$ into our prediction equation, we should obtain the value of $\overline{Y}$.

$$\tilde{Y} = 1.2 + .8\ (3.5) = 1.2 + 2.8 = 4.0 = \overline{Y}$$

Given these numerical values for a and b, we can compute a predicted score for each individual. The difference between an individual's predicted score and his actual score ($Y-\tilde{Y}$) will be the error that we should place in the error vector for that person. For the first person in Figure 6-10, we have a predicted score of 2.0 ($\tilde{Y} = 1.2 + .8\ (1.0) = 2.0$). For this particular subject, the predicted Y value ($\tilde{Y}$) and his actual Y value (Y) are exactly the same, and hence there is no error. The $\tilde{Y}$ value for the second subject is 2.8 ($\tilde{Y} = 1.2 + .8\ (2) = 1.2 + 1.6 = 2.8$). This predicted criterion value is 1.2 units less than the observed criterion value, and hence a value of 1.2 for this subject must be placed in the error vector. The complete linear regression model should now be evident, and can be constructed as in Figure 6-11.

$$Y = aU + bX_1 + E_1$$

$$\begin{array}{ccccccc} Y & & U & & X_1 & & E_1 \\ \begin{bmatrix} 2 \\ 4 \\ 2 \\ 5 \\ 4 \\ 7 \end{bmatrix} & = & 1.2 \begin{bmatrix} 1 \\ 1 \\ 1 \\ 1 \\ 1 \\ 1 \end{bmatrix} & + & .8 \begin{bmatrix} 1 \\ 2 \\ 3 \\ 4 \\ 5 \\ 6 \end{bmatrix} & + & \begin{bmatrix} 0 \\ 1.2 \\ -1.6 \\ +.6 \\ -1.2 \\ +1.0 \end{bmatrix} \end{array}$$

Figure 6-11. Linear regression model with continuous and unit vectors.

We can see in Figure 6-11 the relationship between the equation for a straight line and the linear regression model. The "Y-intercept" is actually the regression weight associated with the unit vector. The "slope" is the regression weight associated with the continuous predictor variable. The components in the error vector are simply the discrepancies between the actual criterion scores and the predicted criterion scores.

Predicting from Continuous and Other Data Combined: Intuitive Approach

In view of the fact that we wish to account for behavior as completely as possible, a combination of predictors can often do a better job than any one predictor alone. This section will present, again at an intuitive level, the use of continuous predictors in conjunction with mutually exclusive predictors.

Consider the data in Figure 6-12. These data are the same as those with which we worked in Figure 6-4.

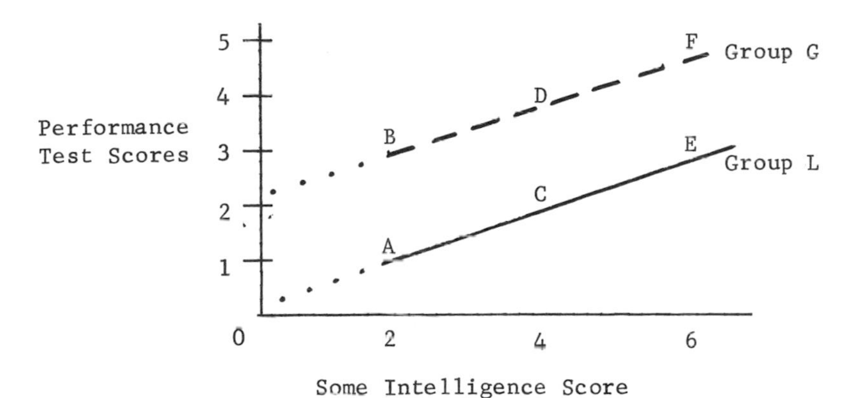

Figure 6-12. Lines which represent the lines of best fit for each group (G and L).

The same six points representing the six students are presented. Within Figure 6-12 two lines were cast. If we know that Students A,

C, and E were taught by Teacher L (lousy) and Students B, D, and F were taught by Teacher G (good), we might ask: "Can the knowledge of group membership (L and G) add to explaining performance when we also have knowledge of intelligence?" As indicated in Figure 6-12, can we cast two lines which further reduce $\Sigma(Y-\tilde{Y})^2$?

It should be obvious that upon construction of weights, etc., we are going to end up with no error.

Box 6.1

We want to reiterate: These examples are provided to give an intuitive "understanding" of model building. In this case you shall shortly see how ridiculous it would be to take this seriously.

From observation of Figure 6-12, you can see that the slopes of the two lines across the X variable are the same (the distance between A and B = 2; C and D = 2; and E and F = 2). In both cases a unit increase in X is accompanied by a .5 increase in Y. Therefore, the weight associated with the variable related to intelligence (the continuous variable) will be .5. This then gives us the slope of both lines.

To complete the equation of the lines, we also need the value of the Y-intercept. The two lines intercept the Y axis at different points: Group L intercepts at 0 and Group G intercepts Y at 2. The task is to reflect these lines in one equation. In a sense, we need different weights to reflect the regression constant. Cast into two equations, we have:

$$\tilde{Y}_G = a_{0_G}U + a_1X_1$$
$$\tilde{Y}_L = a_{0_L}U + a_1X_1$$
where $a_{0_G} \neq a_{0_L}$

Of course, $a_{0_G} = 2$ (the Y-intercept) and $a_{0_L} = 0$ (the Y-intercept). Given: $\tilde{Y}_G = 2U + \frac{1}{2}X$ and $\tilde{Y}_L = 0U + \frac{1}{2}X$, you can see:

$$\Sigma(Y_G - \tilde{Y}_G)^2 = 0$$

and

$$\Sigma(Y_L - \tilde{Y}_L)^2 = 0.$$

If this is not obvious, make two figures like Figure 6-5 -- one for each group -- and prove these conclusions.

Can we place these two equations into one? Let us see.

If we have one Y vector representing both Y_G and Y_L and an X_1 vector representing the appropriate intelligence elements, we only need to provide two vectors to represent the different Y-intercepts.

Let us say:

(6.5a) $$Y_1 = a_0U + a_1X_1 + a_2X_2 + a_3X_3 + E_3$$

where:

X_1 = a vector containing the intelligence scores for each subject.

X_2 = a vector in which the element is 1 if the corresponding score on Y_1 is from a subject who had Teacher L; zero otherwise.

X_3 = a vector in which the element is 1 if the corresponding score on Y_1 is from a subject who had Teacher G; zero otherwise.

and where:

weights a_0, a_1, a_2, and a_3 are calculated to minimize the sum of the squared elements in vector E_3. Figure 6-13 reflects the several vectors.

$$\underset{Y_1}{\begin{bmatrix}1\\3\\2\\4\\3\\5\end{bmatrix}} = a_0 \underset{U}{\begin{bmatrix}1\\1\\1\\1\\1\\1\end{bmatrix}} + a_1 \underset{X_1}{\begin{bmatrix}2\\2\\4\\4\\6\\6\end{bmatrix}} + a_2 \underset{X_2}{\begin{bmatrix}1\\0\\1\\0\\1\\0\end{bmatrix}} + a_3 \underset{X_3}{\begin{bmatrix}0\\1\\0\\1\\0\\1\end{bmatrix}} + \underset{E_3}{\begin{bmatrix}\ \\ \\ \\ \\ \\ \ \end{bmatrix}}$$

Figure 6-13. Vector representation of the unrestricted model given in Equation 6.5a.

From the discussion presented in Chapters III and IV, we know that vector U is linearly dependent upon vectors X_2 and X_3; therefore, one of the weights a_0, a_2, or a_3 is redundant and thus can be set at zero.

We have already shown that $a_1 = .5$. Since the computer program you use always gives the unit vector a weight first, a_2 or a_3 will be set at zero to reflect the linear dependency. If $\underline{a_0}$ is set at $\underline{2}$ (which happens to be the Y-intercept of Group G) and a_2, the weight for Group L, is set at -2, then $a_3 = 0$ (reflecting redundancy). Our equation is:

$$(6.6) \qquad Y_1 = 2U + .5X_1 + -2X_2 + 0X_3 + E_3$$

$$a_0 = 2;\ a_1 = .5;\ a_2 = -2;\ \text{and}\ a_3 = 0.$$

If you solve for E_3 in Figure 6-13 using the above values, the sum of the squared elements in E_3 will equal zero -- errorless prediction as was found with the two separate equations.

Return to our original question: "Can the knowledge of group membership (L and G) add to explaining performance on the achievement test conjointly with knowledge of intelligence?"

Equation 6.6 reflects the unrestricted model (the model with the several sources of knowledge). To answer the question, we must give up knowledge of group membership. This can be done by setting $a_2 = a_3 = a_0$, a common weight.

$$\tilde{Y}_1 = a_0U + a_1X_1$$
and
$$Y_1 = a_0U + a_1X_1 + E_4$$

In essence, the restriction forces the two parallel lines in Figure 6-12 into two superimposed lines as shown in Figure 6-4. The sum of the squared elements in E_4 is $\Sigma(Y-\tilde{Y})^2$, which is equal to 6, presented in Figure 6-4a.

E_4 can be called the ESS_t since we call this the ESS without knowledge; and E_3 can be called the ESS_w since this is the error value calculated with knowledge of group membership:

The F ratio:

$$F = \frac{(ESS_t - ESS_w) / df_1}{ESS_w / df_2}$$

where:

ESS_w = the sum of the squared elements in E_3 = 0

ESS_t = the sum of the squared elements in E_4 = 6

df_1 = 3-2 = 1

df_2 = 6-3 = 3

$$F = \frac{6-0 / 1}{0 / 3} \text{ is undefined.}$$

The F is not "real" because the denominator equals zero; our comment in Box 6.1, on page 128, was given to anticipate this stupidity. Let us arbitrarily add just one bit of error so as not to frustrate the reader. Say E_4 = .1. Then:

$$F = \frac{(6 - .1) / 1}{.1 / 3} = \frac{5.90}{.033} = 178$$

An F of 10.13 is encountered five times in one hundred, and an F of 34.12 is encountered one time in one hundred.

Since an F of 178 is a rare occurrence, even with such small degrees of freedom (1 and 3), the hypothesis that $a_2 = a_3 = a_0$ would be rejected by most investigators.

<u>Problem 6-1: Does the constant difference between lines equal zero?</u>

This problem uses the data given in Example II (Chapter IV) with some added information. Suppose we know the level of test anxiety for each student, and we also know that this level has something to do with the way the students will perform on the test. Our question now is: "Can the knowledge of group membership (Condition 1, 2, or 3) add to explaining achievement on the test conjointly with knowledge of test anxiety?"

These figures illustrate (1) the constant difference between each group's performance; and (2) no difference (without knowledge of group membership).

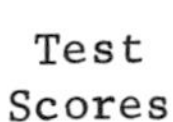

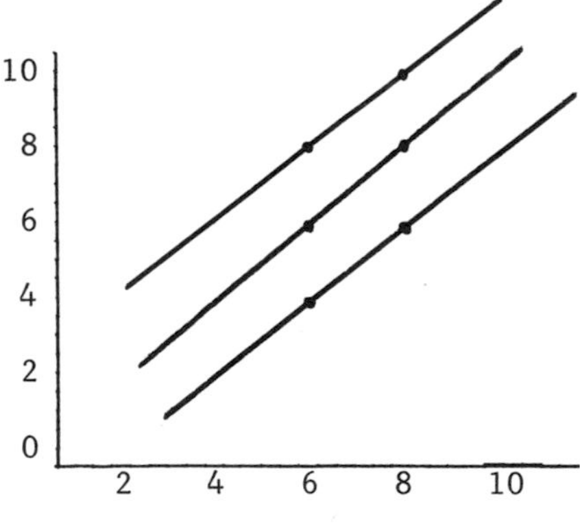

Test Anxiety

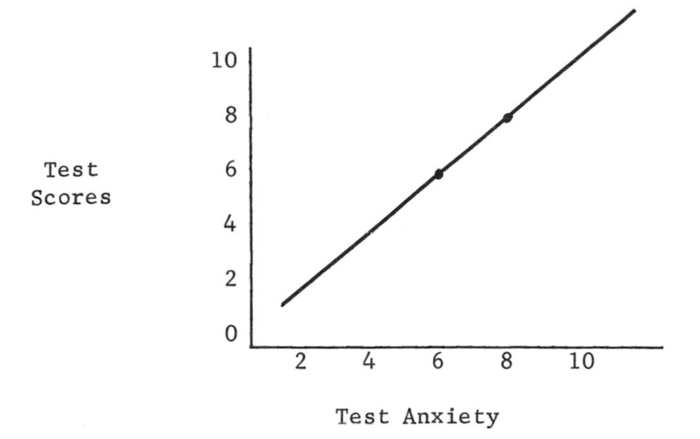

Our question could be stated: "Is the difference between the lines equal to zero?"

This is a listing of the data cards as they now should read:

```
010310004
020510006
030210004
040410005
050810009
060410005
070310004
080910010
090401010
100401006
110301005
120801004
130701004
140401005
150201004
160501009
170600110
180700105
190800106
200600109
210700105
220900104
231000104
240900104
```

The location of the data is as follows:

Columns: 1-2 subject number

3-4 test score (1) Criterion

5	1 if member of Condition 1; zero otherwise.	(2)
6	1 if member of Condition 2; zero otherwise.	(3)
7	1 if member of Condition 3; zero otherwise.	(4)
8-9	test anxiety level	(5)

a. Write a title card, a parameter card, and a format card for this problem. Check your answers with the Answer Section.

b. What is the equation of the unrestricted model? This model should include the knowledge of test anxiety level and of group membership. You may pattern your model after equation 6.5a on page 129. Check with the Answer Section.

c. What is the equation of the restricted model? The restriction necessary to force the three lines into one is: $a_2 = a_3 = a_4 = a_0$. Check your answer.

d. Write model cards for your full and restricted models.

e. Now write an F ratio card that will answer the question. What is the degrees of freedom for Model 1? For Model 2? Remember to count the unit vector as one; with mutually exclusive vectors, one vector is not counted because of linear dependency.

f. Run the problem on the computer with program LINEAR.

Problems with Non-Parallel Slopes: Intuitive Approach

In many psychological studies, the influence of treatment is not the same throughout the range of some concommitant variables. For example, anxiety studies often show that when the treatment is without pressure, high-anxious students perform well; but when the treatment includes great pressure, high-anxious students perform poorly. Figures 6-13a and 6-13b show two cases of what is called "interaction." Figure 6-13a shows the relationship of anxiety and treatment just discussed.

Let us consider Figure 6-13b. We might assume that an investigator wishes to examine the effects of a special program designed to teach grammar in a new abstract format (call this Method A). Method B is the old pedestrian grammar program. The investigator may look at Method A and consider it too difficult for average students; he suspects that the condition expressed in Figure 6-13b exists. He can ask the question in several ways:

1. Is there a constant difference on the grammar achievement test between methods across the observed I.Q. levels?

or

2. Is the influence of method of instruction on achievement different across the observed I.Q. levels?

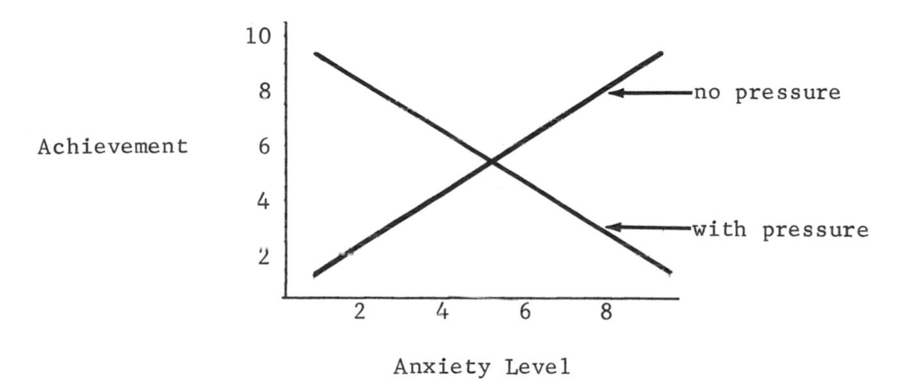

Figure 6-13a. A case where pressure conditions interact with anxiety level.

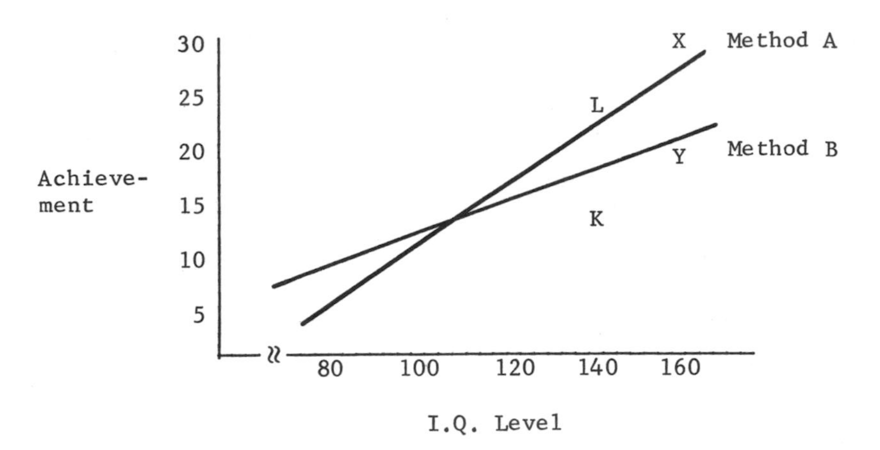

Figure 6-13b. A case where method of instruction interacts with I.Q.

In Question 1, he asks if the difference between L and K is equal to the difference between X and Y (L-K = X-Y). Question 2 phrases the same notion in negative terms (L-K ≠ X-Y). For both questions we must cast the same two models.

If L-K = X-Y, then the slope of both treatments will be equal. If L-K ≠ X-Y, then each treatment must have a different weight associated with I.Q. in order to have two slopes. (Note that in the previous section we considered situations in which the Y-intercepts of the lines cast were unequal, but the slopes of the lines were equal. We now have a situation in which the slopes may also be unequal.)

We can cast a linear model for Method A:

$$\tilde{Y}_1 = a_0U + a_1X_1$$

where X_1 = I.Q. value associated with the observed grammar test score (Y_1).

Method B would be expressed by:

$$\widetilde{Y}_1 = a_2U + a_3X_1.$$

These two equations can be combined to form:

$$Y_1 = a_0U + a_1X_1 + a_2X_2 + a_3X_3 + a_4X_4 + E_5 \quad (6.7)$$

where:

X_1 = 1 if the score on Y_1 comes from a member taught by Method A; zero otherwise.

X_2 = 1 if the score on Y_1 comes from a member taught by Method B; zero otherwise.

X_3 = I.Q. score if the corresponding score on Y_1 comes from the group taught by Method A; zero otherwise.

X_4 = I.Q. score if the corresponding score on Y_1 comes from the group taught by Method B; zero otherwise.

$E_5 = (Y_1 - \widetilde{Y}_1)$.

Equation 6.7 differs from Equation 6.6 in one respect: Both treatments do not share a common weight for the continuous variable; therefore, the slopes of the two lines are <u>not</u> forced to be parallel. Equation 6.7 will <u>allow</u> the lines of best fit to take the form shown in Figure 6-13b.

We can restrict Equation 6.7 to fit the lines so that L-K = X-Y, and then we can see if this fit increases our errors of prediction. This restriction can be expressed as $a_3 = a_4 = a_5$, some common weight.

$$Y_1 = a_0U + a_1X_1 + a_2X_2 + a_5X_3 + a_5X_4 + E_6 \quad (6.8)$$

where all variables are identical with those in Equation 6.7. The vectors will look like those in Figure 6-14.

$$\underset{Y_1}{\begin{bmatrix} Y_{1_1} \\ Y_{1_2} \\ Y_{1_3} \\ Y_{1_4} \\ Y_{1_5} \\ Y_{1_6} \\ \cdot \\ \cdot \\ \cdot \\ Y_{1_n} \end{bmatrix}} = a_0 \underset{U}{\begin{bmatrix} 1 \\ 1 \\ 1 \\ 1 \\ 1 \\ 1 \\ \cdot \\ \cdot \\ \cdot \\ 1 \end{bmatrix}} + a_1 \underset{X_1}{\begin{bmatrix} 1 \\ 1 \\ 1 \\ 0 \\ 0 \\ 0 \\ \cdot \\ \cdot \\ \cdot \\ \cdot \end{bmatrix}} + a_2 \underset{X_2}{\begin{bmatrix} 0 \\ 0 \\ 0 \\ 1 \\ 1 \\ 1 \\ \cdot \\ \cdot \\ \cdot \\ \cdot \end{bmatrix}} + a_3 \underset{X_3}{\begin{bmatrix} 80 \\ 100 \\ 130 \\ 0 \\ 00 \\ 0 \\ \cdot \\ \cdot \\ \cdot \\ \cdot \end{bmatrix}} + a_4 \underset{X_4}{\begin{bmatrix} 0 \\ 0 \\ 0 \\ 80 \\ 100 \\ 140 \\ \cdot \\ \cdot \\ \cdot \\ \cdot \end{bmatrix}} + \underset{E_5}{\begin{bmatrix} \\ \\ \\ \\ \\ \\ \\ \\ \\ \end{bmatrix}}$$

Figure 6-14. The vector form which reflects Equation 6.7

If we look at vectors X_3 and X_4 in Figure 6-14, we can see that if the two vectors are added $(X_3 + X_4)$, we get a continuous vector whose elements are the I.Q. score irrespective of group membership.

Therefore, we can reduce Equation 6.8 to:

(6.9) $$Y_1 = a_0U + a_1X_1 + a_2X_2 + a_5Z_1 + E_6$$

where:

Z_1 = the sum of vectors X_3 and X_4, or a continuous vector whose elements are I.Q. scores regardless of group membership.

a_5 = partial regression weight indicating the slope.

You should note that Equation 6.9 has the same form as Equation 6.4 which provided parallel lines.

If we compute an R^2 using Equation 6.7, the value would indicate the per cent of variance accounted for using group membership separate slopes. This is the $R^2{}_f$.

Likewise, if we compute an R^2 using Equation 6.9, the value would indicate the per cent of variance accounted for using group membership and a common slope. This is the R^2_r.

An F can be calculated using the formula:

$$F = \frac{(R^2_f - R^2_r) / (m_1 - m_2)}{(1 - R^2_f) / (N - m_1)}$$

where:

m_1 = number of weights associated with the number of linearly independent vectors in the full model. Note in Figure 6-14 X_1, X_2, and the vector U are linearly dependent; therefore, one is redundant. m_1, therefore, is 2 plus the two weights associated with X_3 and X_4. $m_1 = 4$.

m_2 = number of weights associated with the number of linearly independent vectors in the restricted model. X_1 and X_2 are also linearly dependent with U; therefore, m_2 = number of weights in Equation 6.9 minus one. $m_2 = 3$.

Problem 6-2: Test for Interaction

Let us provide another test problem for you to run, using program LINEAR. The problem is an extension of the 24N case you have just completed.

Suppose we wish to investigate the influences of test conditions upon performance on a test.

Condition 1: The test result was not to be counted for the course grade.

Condition 2: The test result would account for 90 per cent of the course grade.

Condition 3: The test given was said to be important and would account for 10 per cent of the course grade.

Furthermore, suppose we have been reading Sarason's work, and we might suspect that the level of test anxiety of the student might also be important. In fact, from Sarason's studies we would expect treatment conditions to interact with anxiety such that under Condition 2, the high-anxious student would do poorly and under Condition 1, he would do well. Conversely, the low-anxious should do best under Condition 3, and poorest under Condition 1.

The question might be:

> Is there a constant difference on criterion test performance among the three test conditions across the observed test anxiety scores?

Figure 6-13b illustrates the non-constant difference for two groups, and three parallel lines would illustrate the constant difference situation.

a. Write the model needed to reflect the non-constant difference. It must include three separate weights to reflect the slope across Test Anxiety and three group membership vectors. You can pattern the model after Equation 6.7. Check your model with the model given in the Answer Section.

b. Restrict your full model to force the three lines into a parallel form. Equation 6.9 may help, but try to cast the restriction without it. Check your model with the Answer Section.

Assume you get an R^2_f from (a) and an R^2_r from (b) . Set up the F test, N = 24 (be sure m_1 and m_2 do not include a weight for any linearly dependent vector).

Your data are recorded in the following format: (2X, F2.0, 3F1.0, F2.0).

(2X) columns 1 and 2 are identification numbers 1 through 24.

Vector 1 F2.0 = columns 3 and 4 are the test scores for subjects (criterion, Y_1).

Vector 2 3F1.0 = column 5 = 1 if score on Y_1 comes from a member of Condition 1; zero otherwise.

Vector 3 Column 6 = 1 if score on Y_1 comes from a member of Condition 2; zero otherwise.

Vector 4 Column 7 = 1 if score on Y_1 comes from a member of Condition 3; zero otherwise.

Vector 5 F2.0 = Columns 8 and 9 = anxiety score.

The data are identical to those given in the previous problem section. Your answer to Question "a" should look like this:

$$Y_1 = a_0U + a_1X_1 + a_2X_2 + a_3X_3 + a_4X_4 + a_5X_5 + a_6X_6 + E_1$$

where:

X_1, X_2, and X_3 = mutually exclusive group membership vectors, and

X_4, X_5, and X_6 = the appropriate anxiety score (e.g., X_4 = anxiety score if the score on Y_1 represents a member of Group 1; zero otherwise.)

The data you have punched up from the previous problem have the anxiety scores all in columns 8 and 9. Now you could re-punch these such that:

Columns 8-9 = anxiety score if the score on Y_1 represents a member of Group 1; zero otherwise.

10-11 = anxiety score if the score on Y_1 represents a member of Group 2; zero otherwise.

12-13 = anxiety score if the score on Y_1 represents a member of Group 3; zero otherwise.

DON'T DO IT!

Let us reconsider the matter. Program LINEAR is a bit better than to leave us with such a nasty task. Our Model needs three group membership vectors which we have in vectors 2, 3, and 4. Also we have our criterion in vector 1. The incomplete model card will look like:

MODEL 1 005001.E-050204.. 01

We need three anxiety vectors, one for each treatment. Look at Figure 5-1 (second page of Chapter V). This is a listing of our mainline program. On the fifth line we have the statement SUBROUTINE DATRAN. This means the subroutine will transform data if you give a few instructions. The subroutine reads in your data as indicated in your format and stores it in a location called A(I+...). Then it moves the A(I+...) data to a location, A(J+...). If you read in five vectors, A(I+...) will have five vectors called:

A(I+1); A(I+2); A(I+3); A(I+4); and A(I+5).

In English these are read as: "A sub-I plus one," etc.

Within DATRAN, another subroutine is called from the storage which is MOVCOR (this is on line 12 of the listing in Figure 5-1.) MOVCOR takes A(I+1) and places it intact in location A(J+1). MOVCOR does this for all vectors read in (in this case up to A(J+5). There is room for 100 vectors in location A(J+...). Therefore, if we give a set of instructions between card 10 MOVCOR and card 20 RETURN, we can generate new vectors. In our problem, we need to get three anxiety vectors. We can do this by inserting three cards. The A(J+6) starts in column 7.

```
A(J+6) = A(I+2)*A(I+5)
A(J+7) = A(I+3)*A(I+5)
A(J+8) = A(I+4)*A(I+5)
```

The first card instructs the computer to take each element in vector A(I+2) and multiply it by the corresponding element in A(I+5) and store this product in A(J+6) (a new generated vector). The computer carries this out for each observation (in this case 24 times).

Vector A(J+6) will look like this:

```
04
06
04
05
09
05
04
10
00
00
..
..
..
00    24th observation
```

If you look at your data cards, you will note the first eight subjects have a "1" in column 5 which is A(I+2). All the others have zeros in column 5; vector A(J+6) then is the anxiety vector for subjects under Condition 1. Likewise, A(J+7) = anxiety score if member of Condition 2, and A(J+8) = anxiety score if member of

Condition 3. When you punch these cards, be sure to use the parentheses punches and be sure to start each card in column 7. Place these cards in order between card 10 MOVCOR and 20 RETURN in your program. (These are about the tenth-twelfth cards into your deck, depending upon the number of systems cards your computer center requires.)

We shall return to our friend DATRAN over and over again because it REALLY is useful.

Now we can complete our Model 1 card.

MODEL 1 005001.E-050204060801

Your equation for the restricted model should be:

$$\widetilde{Y}_1 = a_0U + a_1X_1 + a_2X_2 + a_3X_3 + a_7X_7$$

where: $X_7 = X_4 + X_5 + X_6$

You say $a_4 = a_5 = a_6 = a_7$, a common weight.

Model 2 should then be:

MODEL 2 003001.E-05020501

Note: Vector 5 (A(J+5)) = Anxiety Score regardless of group membership.

IMPORTANT NOTE:

You must change your parameter card which was 000240000500005 to 000240000500008. The last value, 8, tells the computer that the original vector plus the generated vectors are eight in number.

c. What should your title, parameter, and format cards look like for this problem?

d. What should your model cards look like?

e. Write your F ratio card and answer the research question.

f. Submit the problem to the Computer Center.

Non-Parallel Slopes: Formal Approach

In Figure 6-13a we find information about data for two treatment conditions: (1) no pressure, and (2) with pressure. The achievement of the subjects was the criterion variable, and the anxiety of the subjects was an independent variable. The following question was asked: "Is the influence of method of instruction (treatments) on achievement (criterion) different across the observed I.Q. levels (independent variable)?" In order to answer this question, it is important to determine if the two lines are parallel. In Figure 6-15a you will find a figure similar to Figure 6-13a with eight specific points.

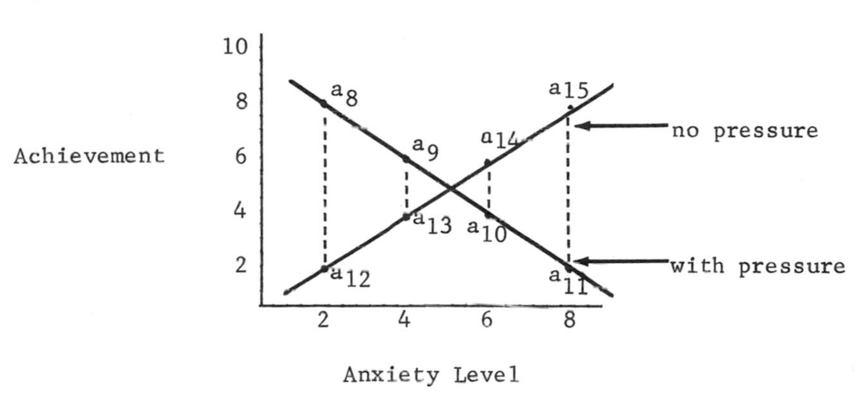

Figure 6-15a. A case where pressure conditions interact with anxiety level.

It is quite typical to find a figure similar to Figure 6-15b in texts using the conventional approach to the problem to illustrate the nature of the data. The eight specific points are cell means of the four anxiety levels on the criterion variable (achievement) for each of the treatment conditions.

		Treatment Condition	
		No Pressure	With Pressure
Anxiety Level	8	a_{15}	a_{11}
	6	a_{14}	a_{10}
	4	a_{13}	a_9
	2	a_{12}	a_8

Figure 6-15 b.

If all cell means (a_i; i = 8, 15) were different and interaction effect was present, knowledge of the cell to which each observation belonged would provide a better estimate of the criterion measure (achievement) than if such information were not used. In order to determine if interaction exists in the data available, the full model would be:

$$Y_1 = a_8X_8 + a_9X_9 + a_{10}X_{10} + a_{11}X_{11} + a_{12}X_{12} + a_{13}X_{13} + a_{14}X_{14} + a_{15}X_{15} +$$

where:

X_8 = a vector in which the element is 1 if the observation is for an individual at the 2 anxiety level taking the pressure treatment; zero otherwise.

X_9 = a vector in which the element is 1 if the observation is for an individual at the 4 anxiety level taking the pressure treatment; zero otherwise.

X_{10} = a vector in which the element is 1 if the observation is for an individual at the 6 anxiety level taking the pressure treatment; zero otherwise.

X_{11} = a vector in which the element is 1 if the observation is for an individual at the 8 anxiety level taking the pressure treatment; zero otherwise.

X_{12} = a vector in which the element is 1 if the observation is for an individual at the 2 anxiety level taking the no-pressure treatment; zero otherwise.

X_{13} = a vector in which the element is 1 if the observation is for an individual at the 4 anxiety level taking the no-pressure treatment; zero otherwise.

X_{14} = a vector in which the element is 1 if the observation is for an individual at the 6 anxiety level taking the no-pressure treatment; zero otherwise.

X_{15} = a vector in which the element is 1 if the observation is for an individual at the 8 anxiety level taking the no-pressure treatment; zero otherwise.

and where:

weights $a_8 \ . \ . \ . \ a_{15}$ are calculated to minimize the sum of the squared elements in the vector E_1.

If the hypothesis of no interaction is to be considered, the following restriction must be made that indicates that the two lines are equidistant from each other:

$$(a_8-a_{12}) = (a_9-a_{13}) = (a_{10}-a_{14}) = (a_{11}-a_{15}) = \text{constant } (K)$$

The above restrictions must be applied to the full model for testing the hypothesis of no interaction. The restrictions are applied by adding and subtracting equal amounts to the full model equation in the following manner:

$$Y_1 = a_8X_8 + (-a_{12}X_8 + a_{12}X_8) + a_9X_9 + (-a_{13}X_9 + a_{13}X_9) + a_{10}X_{10} + (-a_{14}X_{10} + a_{14}X_{10}) + a_{11}X_{11} + (-a_{15}X_{11} + a_{15}X_{11}) + a_{12}X_{12} + a_{13}X_{13} + a_{14}X_{14} + a_{15}X_{15} + E_1$$

By collecting the terms:

$$Y_1 = (a_8 - a_{12})\ X_8 + a_{12}X_8 + (a_9 - a_{13})\ X_9 + a_{13}X_9 + (a_{10} - a_{14})\ X_{10} + a_{14}X_{10} + (a_{11} - a_{15})\ X_{11} + a_{15}X_{11} + a_{12}X_{12} + a_{13}X_{13} + a_{14}X_{14} + a_{15}X_{15} + E_1$$

Assuming no interaction, the following restrictions are considered:

$$(a_8 - a_{12}) = (a_9 - a_{13}) = (a_{10} - a_{14}) = (a_{11} - a_{15}) = \text{constant } (K)$$

Then:

$$Y_1 = K(X_8 + X_9 + X_{10} + X_{11}) + a_{12}(X_8 + X_{12}) + a_{13}(X_9 + X_{13}) + a_{14}(X_{10} + X_{14}) + a_{15}(X_{11} + X_{15}) + E_2$$

But:

$(X_8+X_9+X_{10}+X_{11}) = X_3$ = a vector in which the element is 1 if the observation is for an individual taking the pressure treatment; zero otherwise.

$(X_8+X_{12}) = X_4$ = a vector in which the element is 1 if the observation is for an individual at the 2 anxiety level; zero otherwise.

$(X_9+X_{13}) = X_5$ = a vector in which the element is 1 if the observation is for an individual at the 4 anxiety level; zero otherwise.

$(X_{10}+X_{14}) = X_6$ = a vector in which the element is 1 if the observation is for an individual at the

6 anxiety level; zero otherwise.

$(X_{11}+X_{15}) = X_7$ = a vector in which the element is 1 if the observation is for an individual at the 8 anxiety level; zero otherwise.

Then the model which incorporates the restriction of parallel lines can be stated as follows:

$$Y_1 = KX_3 + a_{12}X_4 + a_{13}X_5 + a_{14}X_6 + a_{15}X_7 + E_2$$

The full and restricted models are compared and the conventional test for interaction is made using the F ratio.

The Purpose for Testing the Significance of the Interaction

The primary purpose of testing for interaction in a problem is to determine if the treatments or conditions are equally effective across all levels of the predictor variable. The nature of interaction may be illustrated by reviewing Figure 6-13a. The results indicate that those students in the pressure treatment group had higher achievement scores than those students in the non-pressure group if the anxiety of the student was less than 4. If the anxiety score was above 4, the students in the non-pressure group performed better than the pressure treatment group. Since the interaction exists, the researcher is _unable_ to make a global statement comparing the two treatment groups across all levels of the predictor variable. (Frequently this global statement is named the "main" effect.) To accurately describe the obtained results, the individual must compare the two treatments at each of the anxiety levels. (Frequently these statements are named "simple" effects.) In the present example, the comparison of the two treatments at each of the four levels of anxiety provides a different result.

If interaction exists in a set of data, the researcher should not attempt to accurately describe the results with a global statement but rather describe the results with statements describing the results at selected levels. If the interaction is found to be absent, an accurate description of the data can be made by comparing the results across all levels. The global statement (or main effect) accurately describes each of the simple effects if interaction does not exist.

The following diagram indicates the steps to be taken when testing for interaction:

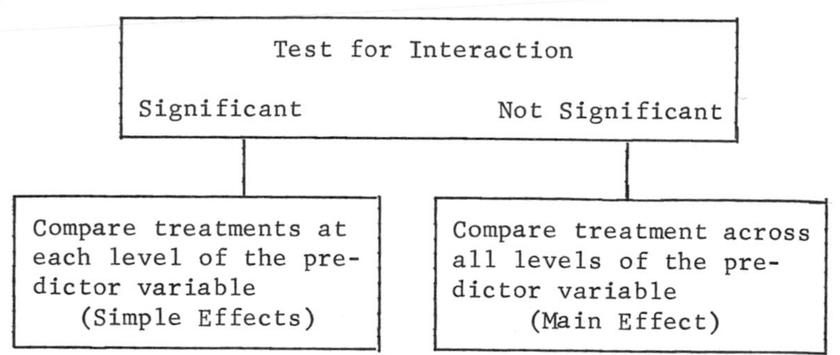

Plotting the Lines of Best Fit

It has been shown that a combination of the weights for the predictor variables, which are printed in the computer output, will give the slopes and Y-intercepts of the lines which represent your models. You now need to know how to use these to plot the graphs of those lines.

Consider the model for Problem 6-1:

$$\widetilde{Y}_1 = a_0U + a_2X_2 + a_3X_3 + a_4X_4 + a_5X_5$$

We know that the unit vector contains a "1" for each subject, that vectors 2, 3, and 4 are mutually exclusive vectors showing group membership, and that vector 5 contains an anxiety score for each subject.

Therefore, for a member of group one, $U = 1$, $X_2 = 1$, $X_3 = 0$, $X_4 = 0$, and X_5 = level of anxiety score. Since vectors 3 and 4 contain zeros for all members of group one, we could rewrite the equation for group one members as:

$$\tilde{Y}_1 = a_0U + a_2X_2 + a_5X_5$$

A similar inspection of the vectors for the model for members of groups two and three would yield equations for group two:

$$\tilde{Y}_1 = a_0U + a_3X_3 + a_5X_5$$

and for group three:

$$\tilde{Y}_1 = a_0U + a_4X_4 + a_5X_5$$

These three equations are all derived from the model and represent the lines of best fit for that model.

To draw a graph of these lines, we look at the output for the problem. The weights for this model are as follows:

Variable Number	Weight
2	0.12344724
3	0.00000000
4	3.12189448
5	0.13220438

Constant = 3.84985206

We want a graph that will show the relationship of anxiety level to test score, so we let the X-axis represent anxiety level and the Y-axis represent test score. We need two points to plot each line, so for each group we may pick any two values for the predictor variable (anxiety level). These can be the same two values for each line. Some investigators choose values one standard deviation on each side of the mean of the predictor; some choose just any high and low values. The predictor values, however, ought to be within the range of the actual scores for that variable.

Suppose we wish to graph the line of best fit for group one. The equation for this line is:

$$\tilde{Y}_1 = a_0U + a_2X_2 + a_5X_5$$

We need two values for X_5 (anxiety level); we may choose the numbers 4 and 6, which fall within the range of scores. We then calculate $\tilde{Y}_1$ for each value of X_5.

(1) If $X_{5_i} = 4$:

$$\tilde{Y}_{1_i} = \begin{cases} a_0U & = 3.85*1 = 3.85 \\ a_2X_{2_i} & = 0.12*1 = 0.12 \\ a_5X_{5_i} & = 0.13*4 = \underline{0.52} \\ & \qquad\qquad\quad 4.49 \end{cases}$$

So for a member of group one with an anxiety level of 4, we would predict a test score of 4.49.

(2) If $X_{5_j} = 6$:

$$\tilde{Y}_{1_j} = \begin{cases} a_0U & = 3.85*1 = 3.85 \\ a_2X_{2_j} & = 0.12*1 = 0.12 \\ a_5X_{5_j} & = 0.13*6 = \underline{0.78} \\ & \qquad\qquad\quad 4.75 \end{cases}$$

This line would be graphed as:

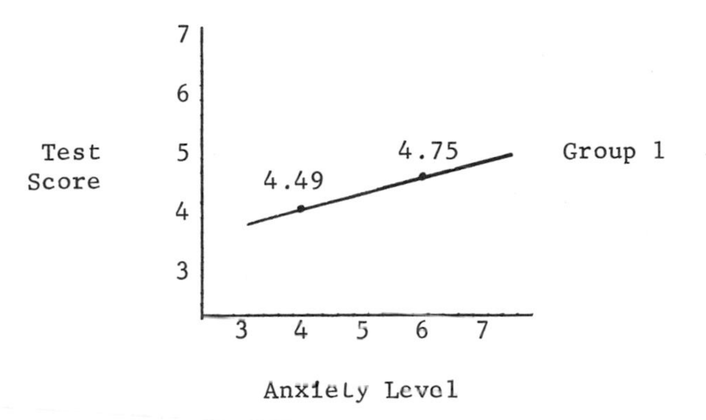

Now let's do the same for group two.

$$\tilde{Y}_1 = a_0U + a_3X_3 + a_5X_5$$

(1) If $X_{5_k} = 4$:

$$\tilde{Y}_{1_k} = \begin{cases} a_0U & = 3.85*1 = 3.85 \\ a_3X_{3_k} & = 0.00*1 = 0.00 \\ a_5X_{5_k} & = 0.13*4 = \underline{0.52} \\ & 4.37 \end{cases}$$

(2) If $X_{5_1} = 6$:

$$\tilde{Y}_{1_1} = \begin{cases} a_0U & = 3.85*1 = 3.85 \\ a_3X_{3_1} & = 0.00*1 = 0.00 \\ a_5X_{5_1} & = 0.13*6 = \underline{0.78} \\ & 4.63 \end{cases}$$

Notice that these two values for Y are each 0.12 less than those for group one. This is equal to the difference between the weights for the group membership variables (0.12 - 0.00).

The graph would now be:

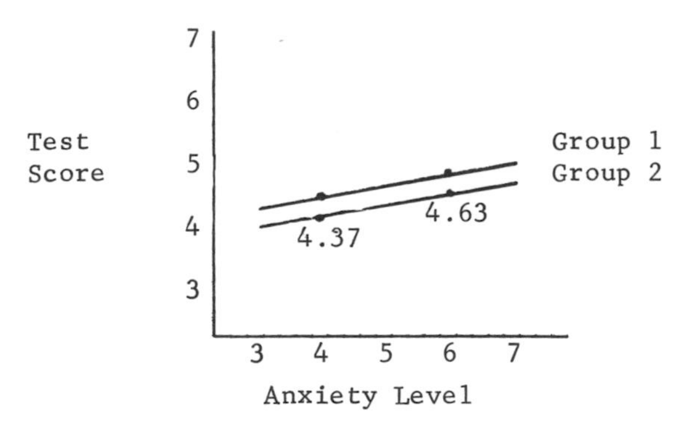

a. Now see if you can add the line for group three. Check your points with those in the Answer Section.

For Further Practice

The restricted model for Problem 6-1 was:

$$\tilde{Y}_1 = a_0U + a_5X_5$$

The output shows these weights:

Variable Number	Weight
5	0.12968917

Constant = 4.94640940

Since we have no knowledge of group membership in this model, there will be only one line of best fit -- that for all subjects combined.

b. To emphasize the point that any two values can be chosen, this time let's use the values 2 and 8 for X_5. Try to plot this line. Check your points with the Answer Section.

Note: For ease of calculation we have derived separate equations for the regression lines representing each group in the model: $\tilde{Y}_1 = a_0U + a_2X_2 + a_3X_3 + a_4X_4 + a_5X_5$. Please note that each vector contains a value for each subject, so each vector actually contributes to the prediction for each line. So we could graph each line using the whole equation, and we would get the same results. For example, we could calculate the same two points for group one, using the complete equation:

(1) If $X_{5_m} = 4$:

$$
\begin{aligned}
& a_0U &= 3.85*1 &= 3.85 \\
\tilde{Y}_{1_m} = \Bigg\{ & a_2X_{2_m} &= 0.12*1 &= 0.12 \\
& a_3X_{3_m} &= 0.00*0 &= 0.00 \\
& a_4X_{4_m} &= 3.12*0 &= 0.00 \\
& a_5X_{5_m} &= 0.13*4 &= \underline{0.52} \\
& & & \ 4.49
\end{aligned}
$$

(2) If $X_{5_n} = 6$:

$$
\begin{aligned}
& a_0U &= 3.85*1 &= 3.85 \\
\tilde{Y}_{1_n} = \Bigg\{ & a_2X_{2_n} &= 0.12*1 &= 0.12 \\
& a_3X_{3_n} &= 0.00*0 &= 0.00 \\
& a_4X_{4_n} &= 3.12*0 &= 0.00 \\
& a_5X_{5_n} &= 0.13*6 &= \underline{0.78} \\
& & & \ 4.75
\end{aligned}
$$

These are the same values obtained when we used the separate derived equation for group one.

Plotting Lines of Interaction

Problem 6-2 contained the full model which allowed interaction between anxiety level and test conditions:

$$\tilde{Y}_1 = a_0U + a_1X_1 + a_2X_2 + a_3X_3 + a_4X_4 + a_5X_5 + a_6X_6$$

where X_1, X_2, and X_3 = mutually exclusive group membership; and where X_4, X_5, and X_6 = the appropriate anxiety score (e.g., X_4 =

anxiety score if the score on Y_1 represents a member of group one; zero otherwise).

We have pointed out that the subscripts may have any value since they have no "real" meaning; they serve only to differentiate between variables. We have found, however, that for purposes of calculation (especially when dealing with the computer output) less confusion is encountered when we subscript each variable according to the order in which it is read by the computer. Therefore, because the predictor variables in the above model correspond to vectors 2, 3, 4, 6, 7, and 8, respectively, we would like to re-write the model as:

$$\tilde{Y}_1 = a_0U + a_2X_2 + a_3X_3 + a_4X_4 + a_6X_6 + a_7X_7 + a_8X_8$$

We will refer to this model in the following discussion.

Just as for Problem 6-1 we will plot a line for each group. Instead of making the lines parallel, this time the slopes of the lines will be allowed to differ so that possible interaction of the data can be assessed. This is done by separating the anxiety scores into different vectors for each group. Thus, each group will now have two weights which are allowed to vary: one for group membership and one for the anxiety scores for that particular group. Let us inspect the model to determine the reduced equation for each line.

A member of group one will have a value of zero in each of vectors X_3, X_4, X_7, and X_8. Therefore, the equation for the line representing the relationship between anxiety level and test score for group one will be:

$$\tilde{Y}_1 = a_0U + a_2X_2 + a_6X_6$$

A member of group two will have a value of zero in each of the vectors X_2, X_4, X_6, and X_8. The equation for the line of group two will be:

$$\tilde{Y}_1 = a_0U + a_3X_3 + a_7X_7$$

A member of group three will have a value of zero in each of the vectors X_2, X_3, X_6, and X_7. The equation for group three will be:

$$\tilde{Y}_1 = a_0U + a_4X_4 + a_8X_8.$$

The output for this problem shows that:

Variable Number	Weight
2	0.00000000
3	6.72435611
4	12.23092031
6	1.04019612
7	-0.12745586
8	-0.53000382

Constant = -1.35961741

Let's plot the line for group one, using the values of 3 and 8 for anxiety. (Remember that we are graphing the relationship of anxiety level to test score for members of group one. In this model, the anxiety score for each group is in a separate vector, and so the calculations for each line will not have a common weight for anxiety as they did for the previous three-line problem.)

(1) If $X_{6_o} = 3$:

$$\tilde{Y}_{1_o} \begin{cases} a_0U &= -1.36*1 = -1.36 \\ a_2X_{2_o} &= 0.00*1 = 0.00 \\ a_6X_{6_o} &= 1.04*3 = \underline{3.12} \end{cases}$$

$$1.76$$

(2) If $X_{6_p} = 8$:

$$\tilde{Y}_{1_p} = \begin{cases} a_0U & = -1.36*1 = -1.36 \\ a_2X_{2_p} & = 0.00*1 = 0.00 \\ a_6X_{6_p} & = 1.04*8 = \underline{8.32} \\ & 6.96 \end{cases}$$

The graph would be:

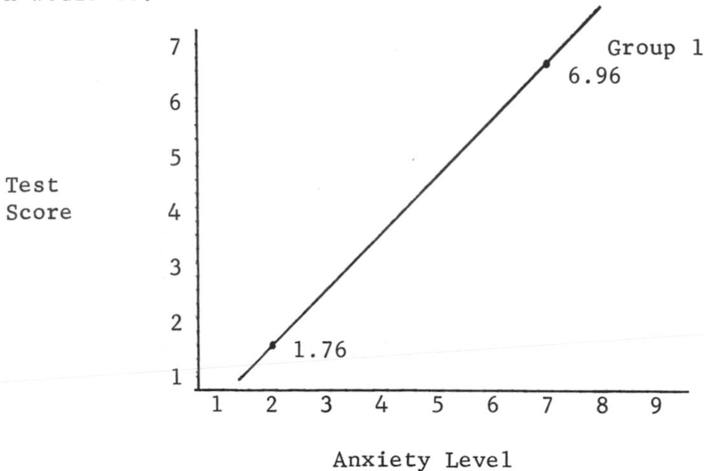

The line for group two, using the same values for anxiety, would be computed as follows:

(1) If $X_{7_q} = 3$:

$$\tilde{Y}_{1_q} = \begin{cases} a_0U & = -1.36*1 = -1.36 \\ a_3X_{3_q} & = 6.72*1 = 6.72 \\ a_7X_{7_q} & = -0.13*3 = \underline{-0.39} \\ & 4.97 \end{cases}$$

(2) If $X_{7_r} = 8$:

$$\tilde{Y}_{1_r} = \begin{cases} a_0U & = -1.36*1 = -1.36 \\ a_3X_{3_r} & = 6.72*1 = 6.72 \\ a_7X_{7_r} & = -0.13*8 = \underline{-1.04} \\ & 4.32 \end{cases}$$

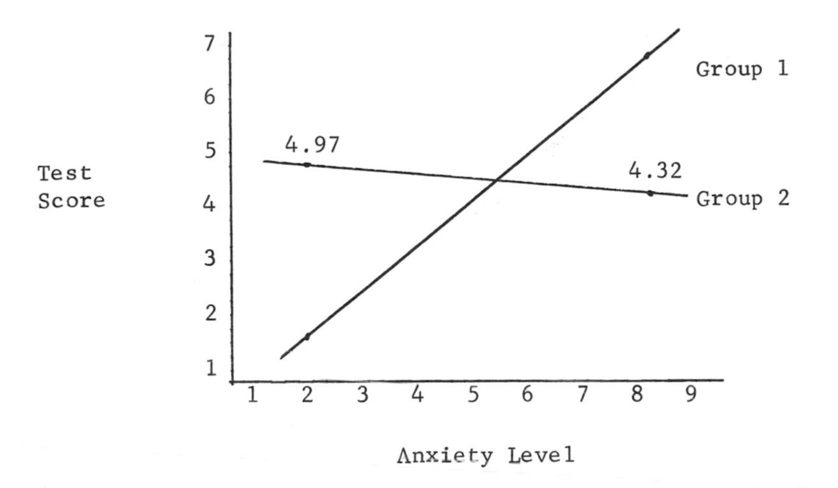

Notice that the lines are definitely not parallel this time. The line for group three will have yet another slope.

Now calculate and add to this graph the line for group three. Check your points with those in the Answer Section.

<u>The Case of Non-Linearity</u>

In the previous section we were using continuous and categorical (mutually exclusive) vectors to predict criterion scores. The models presented took the general form:

(6.15) $$Y_1 = a_0U + a_1X_1 + E_1$$

where: X_1 = a continuous vector.

A representation of this general form is shown in Figure 6-16a.

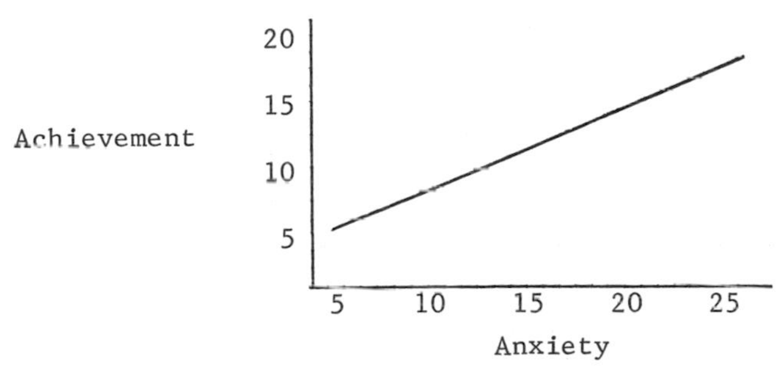

Figure 6-16a. A line representing a possible relationship between X and Y using the general equation $Y_1 = a_0U + a_1X_1 + E_1$.

A line of best fit, however, is not always rectilinear or a series of rectilinear lines. If, for example, moderately-anxious subjects perform best on an achievement task and high- and low-anxious subjects perform less well, we would have a case like the one presented in Figure 6-16b. A number of studies in the field of anxiety research do reveal that a line of best fit for a model answering the research question is non-rectilinear, like Figure 6-16b. You will notice that this figure shows three points rather than an actual <u>line</u>. We cannot cast a model that will reflect such a line. Using previously discussed procedures, we can cast a full model to reflect the <u>points</u>:

(6.16) $$Y_1 = a_0U + a_1X_1 + a_2X_2 + a_3X_3 + E_1$$

where:

X_1 = 1 if the corresponding score on Y_1 comes from an individual who was in the <u>lowest</u> 1/3 on the anxiety measure; zero otherwise.

X_2 = 1 if the corresponding score on Y_1 comes from an individual who was in the <u>middle</u> 1/3 on the anxiety measure; zero otherwise.

X_3 = 1 if the corresponding score on Y_1 comes from an individual who was in the <u>highest</u> 1/3 on the anxiety measure; zero otherwise.

(You should note that a linear dependency exists within the predictor set; can you identify it? If not, check Chapter V.)

We can restrict Equation 6.16 such that $a_1 = a_2 = a_3 = 0$. In this restriction, we are saying that anxiety level is unrelated to achievement. If the F value and the level of significance we set (e.g., $p < .01$; probability of an F this large or larger is less than one time in a hundred) were exceeded by the F ratio for the comparison of the full model with the restricted model, then something in the restriction $a_1 = a_2 = a_3 = 0$, is not true. In the case represented in Figure 6-16b, we probably would find $a_1 = a_3 \neq a_2$. Low and high anxiety subjects perform the same, but moderately-anxious subjects perform in a manner different from the other two. Such a state of affairs has been reported as a curvilinear condition.

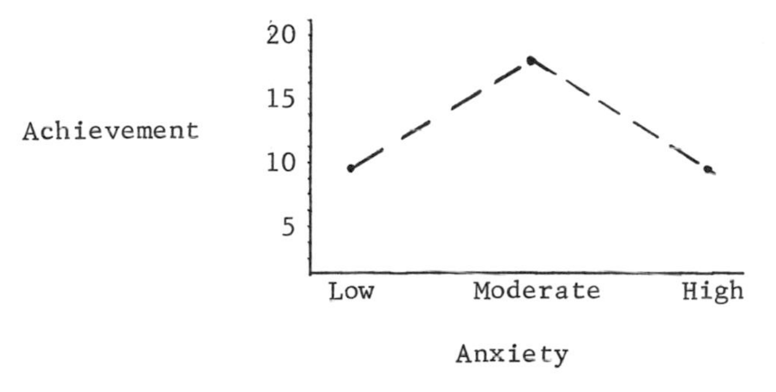

Figure 6-16b. Three points which represent a possible relationship between X and Y using three categories of anxiety rather than the continuous scores and where a curvilinear relation exists.

If Figure 6-16b were smoothed out, we might get a figure something like Figure 6-17a.

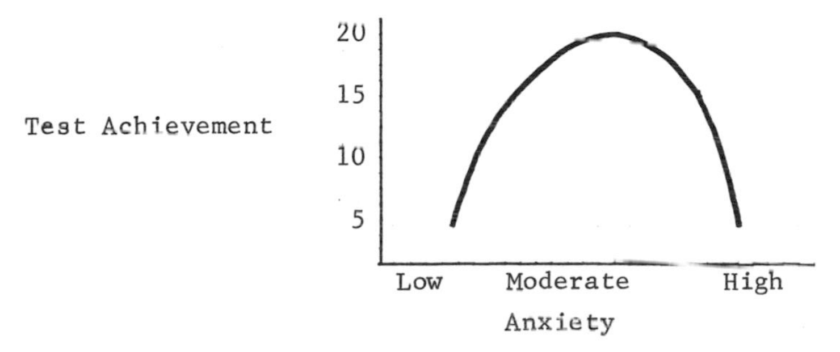

Figure 6-17a. A smoothed-out line that represents Figure 6-16b.

Now, if Equation 6.16 represented the "real" state of affairs, Equation 6.15 would give a larger error term than 6.16. Figure 6-17b shows the probable line (dotted) using Equation 6.15 with the anxiety level-achievement data. The line misses quite often. In a sense, when models include continuous vectors in the predictor set, the general equation $Y_1 = a_0U + a_1X_1 + E_1$ assumes X_1 is rectilinearly related to Y_1 (a straight line). Indeed, if perfect curvilinearity exists in the real data, Equation 6.15 would yield an R^2 of zero and the horizonital line would indicate $a_1 = 0$. A false conclusion might be made (e.g., if $a_1 = 0$, one would conclude the achievement is related to anxiety!).

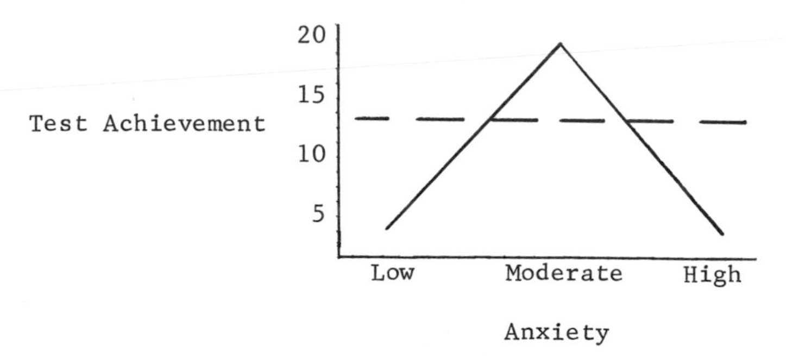

Figure 6-17b. The dotted line would be the line of best fit using equation $Y_1 = a_0U + a_1X_1 + E_1$.

If we expect a curvilinear relationship (from our theory), we could break the continuous vector into three or more categorical vectors and place restrictions to test the curvilinear hypothesis. (Later you will be shown how this can be done using DATRAN, but let us first pursue curvilinear relationships.)

When we have a fairly large group of subjects, say 100 or more, which will probably give us a number of observations at most points on the continuous predictor, categorizing the continuous data (high, moderate, low) will likely decrease the precision of prediction (greater error term) over allowing the actual curved line to exist. For example, if a fairly smooth curvilinear fit (i.e., Figure 6-17a) is real, categorizing will force the X_1 scores at the extreme of each category to have the same $\widetilde{Y}_1$ scores as those who occupy the middle of the same category. If we look at Figure 6-18, it is obvious that the subject with an X_1 score of 4 has a Y_1 score more like the subject with an X_1 score of 5 than the subject with an X_1 score of 2.

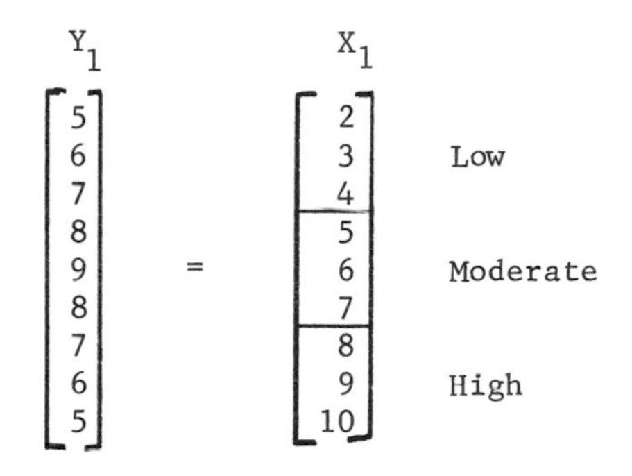

Figure 6-18

The Y_{1_1} for $X_{1_1} = 2$ is 5; the Y_{1_3} for $X_{1_3} = 4$ is 7; and the Y_{1_4} for $X_{1_4} = 5$ is 8. Nevertheless, if we break X_1 into three groups, low, moderate, and high, $X_{1_1} = 2$ and $X_{1_3} = 4$ are placed in the same group and $X_{1_4} = 5$ is placed in another. When we have a large number of observations, such categorizing will yield an increase in the error.

We can overcome this loss by retaining the continuous vector and generate another vector which will allow the line to curve. This can be done by adding a vector whose elements are the squares of the elements in the continuous vector.

(6.17) $$Y_1 = a_0U + a_1X_1 + a_2X_1^2 + E_1$$

where:

Y_1 = criterion score.

X_1 = a continuous vector of anxiety scores corresponding to the subject's Y_1 score.

X_1^2 = a continuous vector of squared anxiety scores corresponding to the subject's Y_1 score.

E_1 = $(Y_1 - \widetilde{Y}_1)$.

a_0, a_1, and a_2 are partial regression weights calculated so as to minimize E_1.

One may ask: "Is there a curvilinear relationship between the achievement and the measured anxiety?" Equation 6.17 can be the full model and Equation 6.18 can be cast to restrict $a_2 = 0$, which then takes $a_2X_1^2$ out of the equation.

(6.18) $$Y_1 = a_0U + a_1X_1 + E_2$$

And the F test can be run to determine the tenability of the restriction. The curvilinear relationship may take many forms, depending upon the magnitude of and signs associated with weights a_0, a_1, and a_2.

A few general observations can be made from the set of Figure 6-19:

1. When a_2 is negative, the open end of the curve will be down; when a_2 is positive, the open end is up.
2. When a_1 and a_2 have like signs, the point of inflection is displaced to the left. When a_1 and a_2 have unlike signs, the displacement is to the right.
3. When $a_1 = 0$, the curve will be symmetrical around the Y axis.
4. a_0 gives the point of Y-intercept (when the model reflects only one line).

(a)

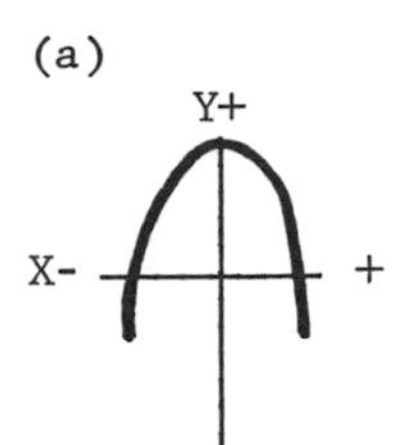

$a_0 > 0$
$a_1 = 0$
a_2 = negative value

(b)

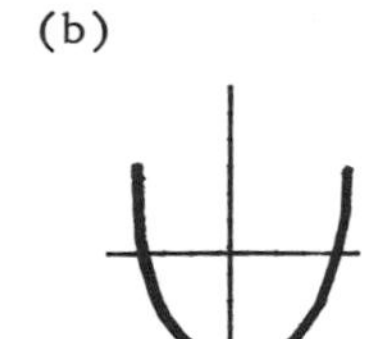

$a_0 < 0$
$a_1 = 0$
a_2 = positive value

(c)

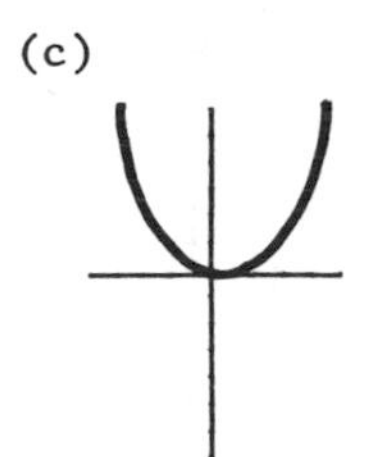

$a_0 = 0$
$a_1 = 0$
a_2 = positive value

(d)

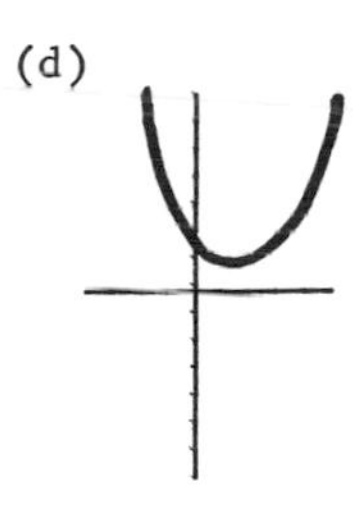

$a_0 > 0$
a_1 = negative value
a_2 = positive value

(e)

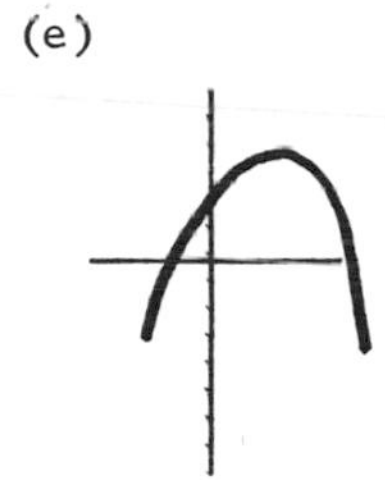

$a_0 > 0$
a_1 = positive value
a_2 = negative value

(f)

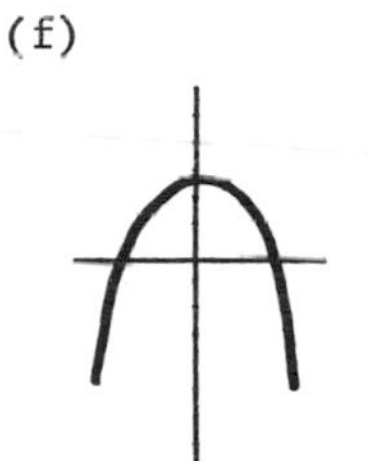

$a_0 > 0$
a_1 = negative value
a_2 = negative value

(g)

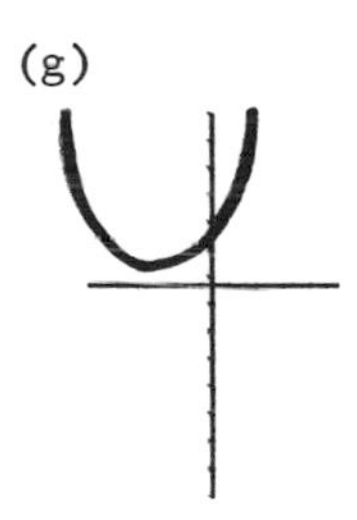

$a_0 > 0$
a_1 = positive value
a_2 = positive value

Figure 6-19

The severity of the curve depends upon the absolute weight of a_2; the larger the weight, the sharper the curve. If we take the values 2, 4, 8, and 6 and set $a_0 = a_1 = 0$ and $a_2 \neq 0$, we can demonstrate this. See Figure 6-20a and Figure 6-20b.

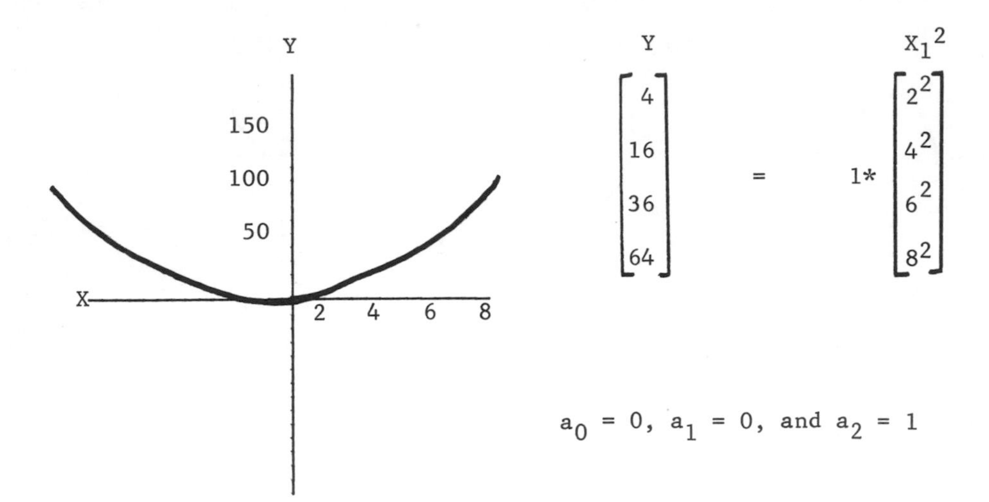

Figure 6-20a

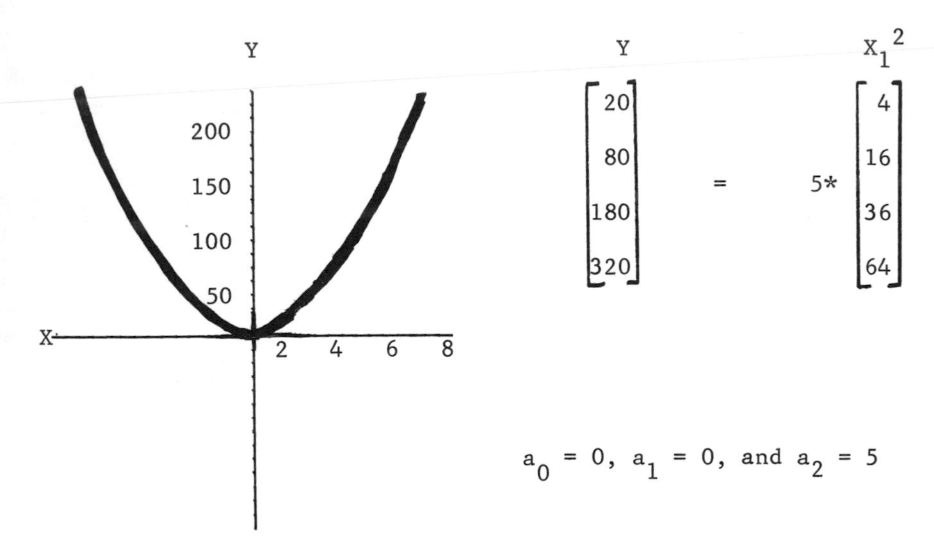

Figure 6-20b

Many of the figures in set 6-19 use all four quadrants. If Y is a change score (e.g., a post-test minus a delayed post-test), one might expect positive and negative values, thus requiring positive and negative representations of Y. In behavioral research we seldom use a positive and negative X value; therefore, the two quadrants on the left are seldom used. In fact, most often both X and Y values are typically positive in behavioral research so that most curves we encounter will be in the upper right-hand quadrant.

Equation 6.17 ($Y_1 = a_0U + a_1X_1 + a_2X_1{}^2 + E_1$) can be developed within program LINEAR through the use of DATRAN. If the criterion (Y_1) was the first variable on the card and the anxiety (X_1) second, we can get X_1^2 by punching a card:

```
A(J+3) = A(I+2)*A(J+2)
                          or
A(J+3) = A(I+2)**2.0
```

(Remember the letter "A" starts in column 7.) The asterisk (*) means multiplication. A(I+2)*A(I+2) gives the new vector A(J+3) whose elements will be the squared elements in A(I+2).

> Remember that the form A(I+P) refers to a variable read by the format card, where P is the ordinal number of that vector (2 if it is the second vector). A(J+Q) refers to a variable produced by a DATRAN statement, where Q is the ordinal number of that vector. The first Q is the number following the last P. For example, if you have read 5 vectors with your format card and if you wish to square the elements in the fourth vector, your transformation card would read:
>
> A(J+6) = A(I+4)**2.
>
> If you then wish to multiply the elements in the third vector by the elements in the fourth vector, your transformation card would read:
>
> A(J+7) = A(I+3)*A(I+4)

If we get a line similar to Figure 6-17a using Equation 6.17, then: a_2 will have a negative weight (open end down) and a_1 will have a positive weight (the point of inflection is to the right of Y).

Other Non-Linear Forms

A. Third-Degree Polynomials with a Plateau in the Middle Range. The case where the curve changes direction once and is symmetrical is only one possible curvilinear form. Indeed, in many psychomotor tasks (e.g., typing) where practice is the X variable and correct responses the Y variable, one may find a plateau as illustrated in Figure 6-21a.

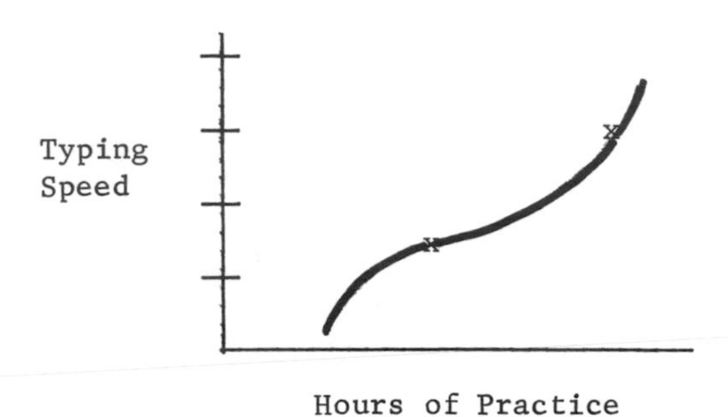

Hours of Practice

Figure 6-21a

An equation which can reflect the two points of inflection would include a third-degree polynomial.

(6.19) $$Y_1 = a_0U + a_1X_1 + a_2X_1^2 + a_3X_1^3 + E_2$$

Equation 6.19 includes the vector whose elements are the cubed elements of X_1. DATRAN statements can provide the generated vector:

```
A(J+4) = A(I+2)**3.0
                          or
A(J+4) = A(I+2)*A(J+3)
```

Since A(J+3) contains the squared elements of A(I+2), the multiplication of these two yields the same vector as does exponentiation.

B. Third-Degree Polynomials with a Plateau in the Extreme Ranges. Due to ceiling and floor effects of some criterion measures, the investigator may expect an S-shaped curve where the plateau is found in the extremes. (See Figure 6-21b.) Such a curve can also be approximated by using the third-degree polynomial.

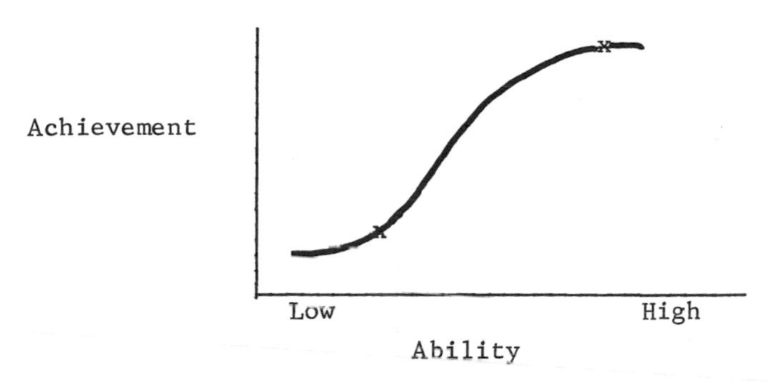

Figure 6-21b. An S-shaped curve with a plateau in the extreme ranges.

The appropriate equation to reflect this form is the same as given in Equation 6.19. Indeed, to determine the shape of the curve, you must plot several points using the appropriate partial regression weights.

Problem 6-3

We have collected data on sixty subjects, and it has been arranged on data cards in the following manner:

columns 1-2: criterion variable
3-4: predictor variable

The cards are punched as follows:

0302
0302
0402
0402
0503
0503
0503
0603
0603
0707
0605
0604

0604
0604
0604
0504
0704
0605
0605
0605
0705
0606
0606
0606
0606
0706
0706
0607
0607
0607
1212
0607
0607
0707
0707
0608
0608
0608
0708
0708
0708
0609
0609
0709
0709
0709
0709
0710
0710
0810
0810
0810
0810
0911
0911
1011
1011
1112
1212
1212

From a study of previous research in this area, we suspect that the relationship between these two variables is a curved one, and possibly contains two points of inflection. To test for this, we will need to square and to cube the predictor variable.

a. Write the transformation cards needed to generate these two vectors.

b. What is the full model that will allow the relationship to be reflected by a line with two inflection points?

c. What is the restriction that will allow only one point of inflection? Write the resulting restricted model.

d. If we find that a_4 does equal zero, we will want to know if there is indeed a curved relationship. What restriction will now force the lines of relationship to be straight? Write the resulting restricted model.

e. Write the model cards for these three models and F ratio cards to test the difference in errors of prediction between the first and second and between the second and third models.

f. Now write the title card, parameter card, and format card necessary to run this problem on the computer.

g. Run the problem. What is the answer to the question?

<u>Problem 6-4</u>

We have data on cards for sixty subjects in the following order:

columns 1-2: criterion variable
3-4: predictor variable

The cards are punched as follows:

0102
0102
0102
0202
0103
0103
0103
0203
0104
0104
0104
0204
0204
0204
0204
0205
0205
0205
0305
0305
0305
0305
0305
0406
0406
0406
0406
0506
0306
0507
0507
0507
0507
0607
0607
0607
0708
0708
0808
0808
0808
0808
0909
0909
0909
0909
0809
1009
0910
0910

0910
1010
0911
0911
0911
1011
0912
0912
0912
1012

This time we expect an S-shaped curved relationship with a plateau in the extremes. We want to ask the question: "Is the relationship between these two variables reflected by an S-shaped curve?"

See if you can set up this problem and run it yourself. You will need:

2 transformation cards

statistical equations

model cards

F ratio cards

title, parameter, and format cards

What is the answer to the question?

Curvilinear Interaction

Just as we combined continuous and dichotomous vectors in rectilinear forms, we can combine curved vectors in the same manner.

Consider the equation for a line with one point of inflection:

$$\tilde{Y}_1 = a_0U + a_1X_1 + a_2X_1^2$$

If we had two groups that are different on some third variable, we can cast a curved line for each:

$$\tilde{Y}_1 = a_0U + a_1X_1 + a_2X_1^2 \quad \text{and}$$
$$\tilde{Y}_1 = a_3U + a_4X_1 + a_5X_1^2$$

Furthermore, we can cast these two equations into one:

(6.21) $$Y_1 = a_0U + a_1X_1 + a_2X_2 + a_3X_3 + a_4X_4 + a_5X_5 + a_6X_6 + E_1$$

where:

X_1 = 1 if the score on Y_1 comes from a member of Group 1; zero otherwise.

X_2 = 1 if the score on Y_1 comes from a member of Group 2; zero otherwise.

X_3 = the score on the continuous variable if the score on Y_1 comes from a member of Group 1; zero otherwise.

X_4 = the score on the continuous variable if the score on Y_1 comes from a member of Group 2; zero otherwise.

X_5 = the square of the elements in X_3.

X_6 = the square of the elements in X_4.

For illustrative purposes, let us define the several variables as:

Y_1 = a leadership score obtained from a peer nomination form.

X_1 = 1 if the student was nominated as conforming; zero otherwise.

X_2 = 1 if the student was nominated as non-conforming; zero otherwise.

X_3 = authoritarian score for conformers; zero otherwise.

X_4 = authoritarian score for non-conformers; zero otherwise.

X_5 = squared elements in X_3.

X_6 = squared elements in X_4.

We might ask:

Question 1: Is the influence of conformity group on leadership non-linear and different across the observed authoritarian scores?

Since equations such as 6.21 use up a large number of the degrees of freedom, the first restriction might ask: Is there any significant predictive variance in the full model (is the RSQ_f significantly different from zero)? The restriction $a_1 = a_2 = a_3 = a_4 = a_5 = a_6 = a_0$ can give us the equation:

(6.22) $$Y_1 = a_0U + E_2$$

The F ratio would be:

$$F = \frac{(RSQ_f - 0) / (6 - 1)}{(1 - RSQ_f) / (N - 6)}$$

Note: $m_1 = 6$ since U, X_1, and X_2 are linearly dependent.

If the F ratio does not satisfy your pre-determined alpha level, you go back to the drawing board (your theory has not been productive).

On the other hand, if the F ratio does satisfy your alpha level, a number of predictive possibilities exist to account for the loss using Equation 6.22 ($Y_1 = a_0U + E_2$). These possibilities follow two trends: (1) where the curved interaction is rejected; and (2) where the curved interaction is accepted.

Trend I: The Test of Possible Rectilinear Forms. The four figures A, B, B_1, and B_2 in Figure 6-23 show the testing order which this section will follow.

$$Y_1 = a_0U + a_1X_1 + a_2X_2 + a_3X_3 + a_4X_4 + a_5X_5 + a_6X_6 + E_1 \quad \text{(Model 1)}$$

Hypothesize $a_1 = a_2 = a_3 = a_4 = a_5 = a_6 = 0$

Is the RSQ_f probably 0? If yes, nice try! Go back to your theory.

If no, Model 1 might yield Figure A:

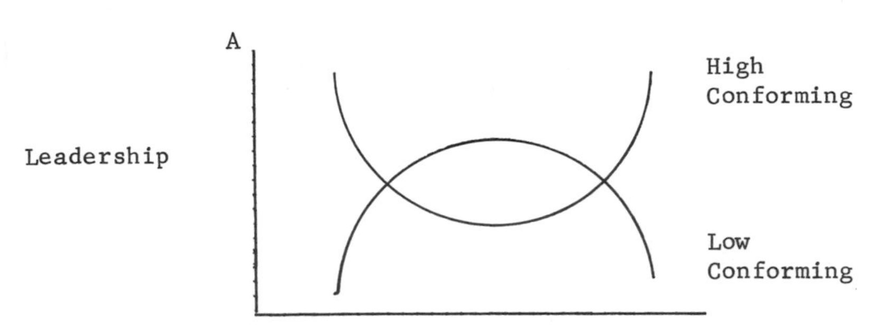

Hypothesize $a_5 = a_6 = 0$

$$Y_1 = a_0U + a_1X_1 + a_2X_2 + a_3X_3 + a_4X_4 + E_2 \quad \text{(Model 2)}$$

If YES ⇩ If NO ⇨ Go to Figure 6-24.

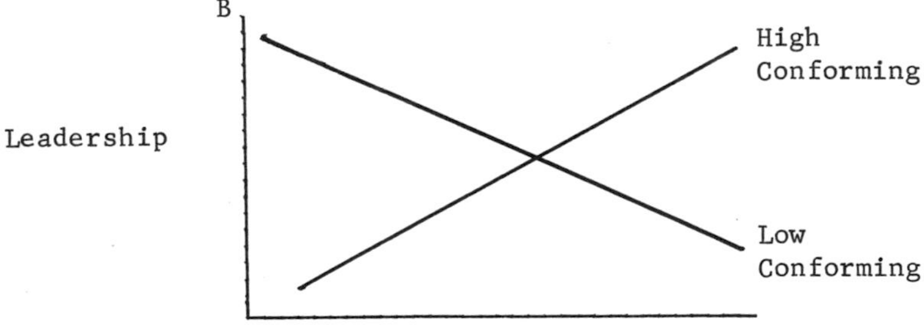

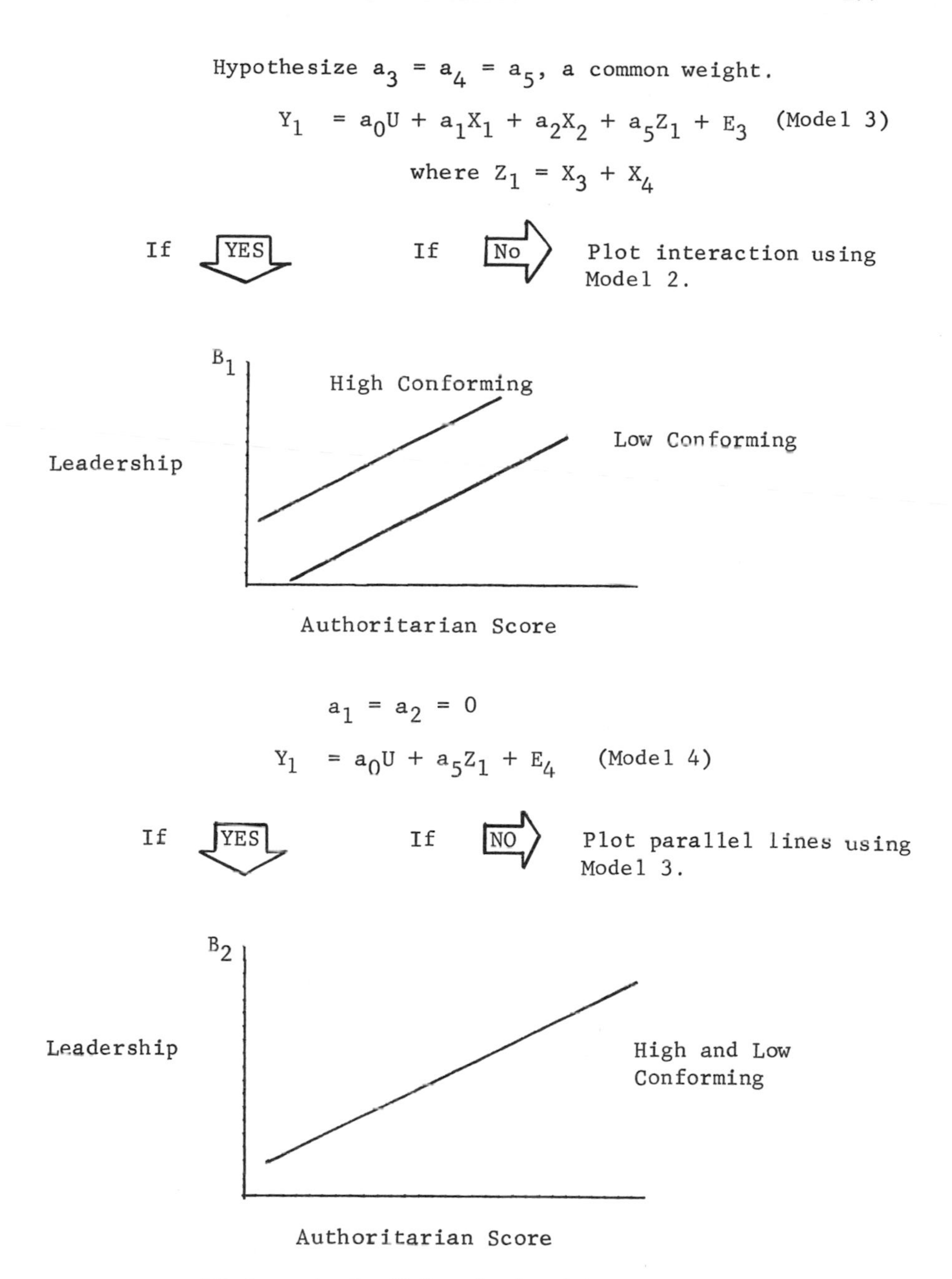

Figure 6-23. Tests for relationships if curvilinear relations are ruled out.

<u>Question 2</u>: Suppose we do not expect a curvilinear interaction but want to test the assumption rather than leave it as an assumption. Equation 6.21 reflects the non-linear interaction expressed in Question 1. To test the question, "Is the assumption correct that there is a linear relationship between authoritarian score and leadership score?", we impose the restriction $a_5 = a_6 = 0$. If $a_5 = a_6$ and are equal to zero, then we have ruled out curved parallel and curved super-imposed lines. The two equations are:

(6.23) (Model 1) $Y_1 = a_0U + a_1X_1 + a_2X_2 + a_3X_3 + a_4X_4 + a_5X_5 + a_6X_6 + E_1$

(Model 2) $Y_1 = a_0U + a_1X_1 + a_2X_2 + a_3X_3 + a_4X_4 + E_2$

If Model 2 does not increase our errors of prediction (if the F ratio between the full and restricted models does not satisfy your alpha level), then we know that the lines are not curved. The restricted model (Model 2) allows the lines to be non-parallel; but we still have two other possibilities: parallel lines (B_1 - Figure 6-23) or superimposed lines (B_2). If Model 2 does not increase errors, we might ask:

<u>Question 3</u>: Is the influence of conformity on leadership score constant across the observed authoritarian measure?

Model 2 is the new full model and the restriction imposed is: $a_3 = a_4 = a_5$, a common weight:

(6.24) (Model 3) $Y_1 = a_0U + a_1X_1 + a_2X_2 + a_5Z_1 + E_3$

where: $Z_1 = X_3 + X_4$.

Note we are forcing the lines to be parallel since both groups are sharing the same weight for the authoritarian measure. If $a_3 \neq a_4 \neq a_5$, the interaction case is supported, and you should plot the interacting lines. If $a_3 = a_4 = a_5$, a common weight, then the interaction is ruled out. Therefore, you might ask:

Question 4: Is the influence of conformity group on leadership equal (or the same) across the observed range of observed authoritarian scores?

Equation 6.24 is the new full model and the restriction is imposed: $a_1 = a_2 = a_0$.

(6.25) (Model 4) $Y_1 = a_0U + a_1Z_1 + E_4$

If $a_1 \neq a_2 \neq a_0$, then parallel lines of some sort exist (e.g., B_1 in Figure 6-23), and a plot of the lines will reveal the relationship. If $a_1 = a_2 = a_0$, then one line, the simple rectilinear X-Y relationship, best describes the data.

Look at Figure 6-23. We have gone from 6-23A to a zero relationship, and then to 6-23B, 6-23B_1, and 6-23B_2.

You may note that Figure 6-23B_2 is the first type of linear situation we discussed early in this chapter. Then we dealt with parallel lines as in 6-23B_1, interacting lines as in 6-23B, and finally with curved lines as in 6-23A. We have discussed them previously in this order, hoping to follow a progression from least to most complex. In a research situation, however, the actual questions will be more likely to progress in the order shown in this section, especially as summarized in Figure 6-23 and Figure 6-24.

<u>Trend II: The Test of Possible Curvilinear Forms</u>. In the case where $a_5 \neq a_6 \neq 0$ (therefore, the lines are curved), we now have Figure 6-24A, B_1, B_2, and B_3 (see Figure 6-24).

(Model 1) $Y_1 = a_0U + a_1X_1 + a_2X_2 + a_3X_3 + a_4X_4 + a_5X_5 + a_6X_6 + E_1$

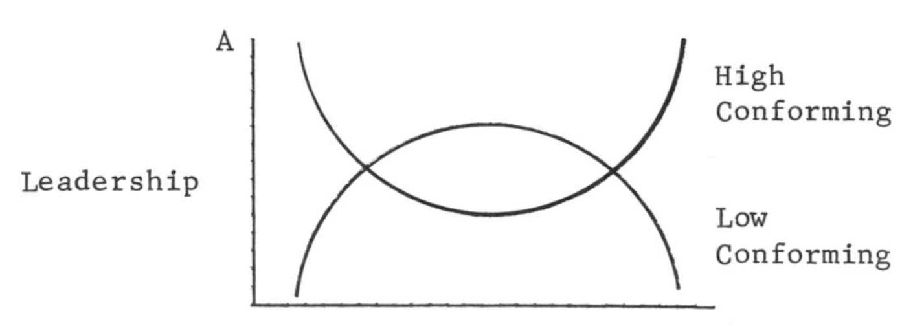

Authoritarian Score

Hypothesize: $a_3 = a_4 = a_7$ (a common weight) and $a_5 = a_6 = a_8$ (a common weight).

(Model 5) $Y_1 = a_0U + a_1X_1 + a_2X_2 + a_7Z_2 + a_8Z_3 + E_5$

where: $Z_2 = X_3 + X_4$ and $Z_3 = X_5 + X_6$

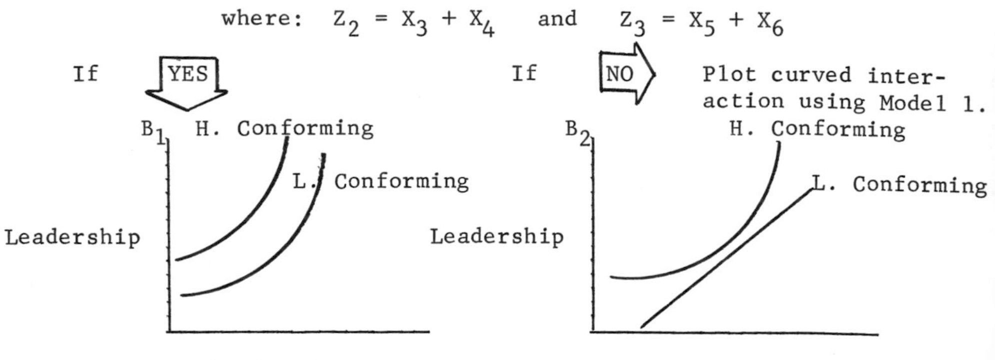

Hypothesize: $a_1 = a_2 = a_0$.

(Model 6) $Y_1 = a_0U + a_7Z_2 + a_8Z_3 + E_6$.

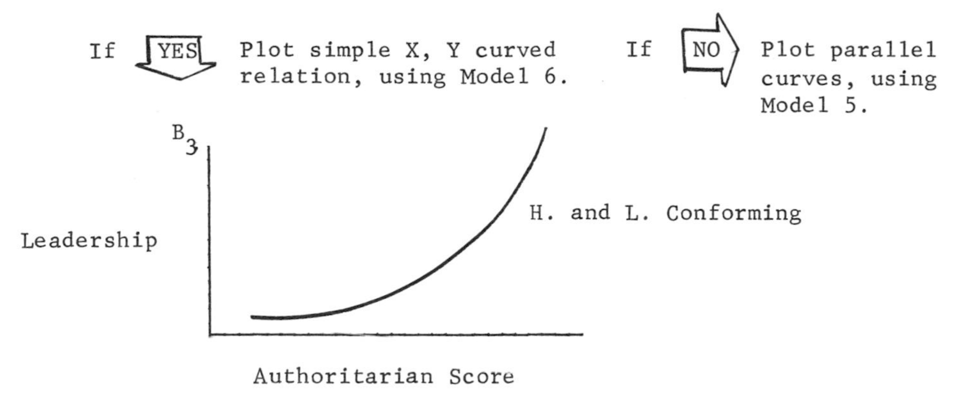

Figure 6-24.

We can restrict the full model (Model 1) such that:

$a_3 = a_4 = a_7$, a common weight, and $a_5 = a_6 = a_8$, a common weight.

(6.26) (Model 5) $Y_1 = a_0U + a_1X_1 + a_2X_2 + a_7Z_2 + a_8Z_3 + E_5$

where: $Z_2 = X_3 + X_4$ and $Z_3 = X_5 + X_6$

This restriction will answer the question:

<u>Question 5</u>: Is the non-linear influence of conformity on group membership constant across the observed authoritarian measure? (Are the curved lines all equidistant from one another at all authoritarian levels?)

If $a_3 \neq a_4$ and $a_5 \neq a_6$ (the ratio is not distributed as an F), then the curvilinear interaction is the state of affairs and we can plot the interaction which might look like 6-24A. If, on the other hand, the figure looks like 6-23B_2, then a_5 or a_6 (which ever weight represents the visually non-curved line) might equal zero. If one line looks rectilinear, you might wish to test the linearity by imposing the restriction ($a_5 = 0$ or $a_6 = 0$, which ever is the appropriate weight).

If $a_3 = a_4$ and $a_5 = a_6$, we might ask Question 6:

Question 6: Is the non-linear influence of conformity group membership equal across the observed authoritarian measure?

The restriction is $a_1 = a_2 = a_0$ and Equation 6.27 reflects the restriction.

(6.27) (Model 6) $Y_1 = a_0U + a_7Z_2 + a_8Z_3 + E_6$

If $a_1 \neq a_2 \neq a_0$, then parallel curved lines represent the lines of best fit. If $a_1 = a_2 = a_0$, then one curved line is the best fit.

This section on curvilinear interaction has explored the possible predictive equations which may reflect lines of best fit and the order in which they might be considered. If an investigator has reason to believe that he is dealing with variables which may be curvilinearly related (or wishes to test the assumption that they are rectilinearly related), he will want to cast an equation containing second- and possibly third-degree polynomials.

> These models do not force a line to be curved; they merely allow it to be curved if that is the best fit for the data. This is the general case of all full and restricted model relationships; the full allows a state of affairs and the restricted does not allow it. If the F ratio calculated between those two models is of significant magnitude, then you know that the best fit is obtained by the full model.

The investigator will then place a restriction on the model such that the line(s) will not be allowed to curve. This usually means setting the weights of the squared-term vectors equal to each other and to zero, effectively dropping them out of the equation.

If that restriction does not increase the errors of prediction (and thereby decrease R^2) significantly, then the investigator will pursue the testing of possible rectilinear forms: interacting straight lines, parallel lines, superimposed straight lines (one line). See Figure 6-23. He will plot the graph of the model which gives the best fit.

If, however, the restriction that did not allow the lines to curve does increase the errors of prediction, the investigator will pursue the testing of possible curvilinear forms: interacting curved lines, parallel curved lines, superimposed curved lines (one line). See Figure 6-24. He will plot the graph of the model which gives the best fit.

Comments Regarding Notation

The notation one uses to cast linear models should be convenient and unambiguous to the user. Some investigators find no difficulty in changing from one set of subscripting in the linear equation to another set of notation for the program LINEAR (when he calls the variables for the models on the model card). The author of this section tends to use various sets depending upon his mood. Some students find such change difficult to follow. You may wish to follow some order, and the comments below are intended to suggest one mode.

Suppose you plan to investigate the curvilinear interaction state of affairs given in the preceding problem section. The data were organized on the card such that:

Variable 1	Col 1-2	= Criterion
2	3	= Group 1
3	4	= Group 2
4	5	= A continuous vector

These data are stored and numbered in sequence of the read-in 1, 2, 3, 4. You had to generate four vectors:

```
A(J+5) = A(I+2)*A(I+4)
A(J+6) = A(I+3)*A(I+4)
A(J+7) = A(J+5)*A(J+5)
A(J+8) = A(J+6)*A(J+6)
```

When you know the order of the vectors to be used on the computer, you might wish to construct your linear equations using those numbers for subscripting since your output will have these numbers. For example, in the full interacting model:

$$Y_1 = a_0U + a_2X_2 + a_3X_3 + a_5X_5 + a_6X_6 + a_7X_7 + a_8X_8 + E_1$$

X_1 was not used because the criterion is vector 1 in the computer; X_4 was not used since X_6 and X_7 are linearly dependent with X_4.

If you hypothesize that there is a rectilinear interaction, the restriction would be $a_7 = a_8 = 0$. The restricted model would be:

$$Y_1 = a_0U + a_2X_2 + a_3X_3 + a_5X_5 + a_6X_6 + E_2$$

The restriction to force the two lines to be parallel might be $a_5 = a_6 = a_9$, a common weight. However, X_4 will give the common slope that would be given by a_9X_9; therefore, to reduce confusion you might say $a_5 = a_6 = a_4$, and your restricted model would be:

$$Y_1 = a_0U + a_2X_2 + a_3X_3 + a_4X_4 + E_3$$

This process of subscripting is identical to the computer numbering system and can reduce silly errors, especially for the novice. We recommend, however, that you abandon the crutch when you become more familiar with the two systems since such usage can make the process a mechanical cookbook act. The intent of this book is to free you from a non-understanding "cookbook" approach to the formulation of research models.

Problem 6-5

From the theoretical study with which we have dealt on the last several pages, we have collected data from and punched cards for sixty subjects. These cards contain information as follows:

columns 1-2 = leadership measures; criterion score.
3 = 1 if low conforming; zero otherwise.
4 = 1 if high conforming; zero otherwise.
5-6 = authoritarian score.

The cards are punched as follows:

011001
011001
011002
021002
021002
021003
031003
041003
051004
061004
071005
061005
061005
071005
071005
061006
061006
061006
051004
080102

071006
051007
051007
061007
041008
041008
021009
021009
031009
021010
011010
100101
090101
090102
080102
070102
070103
070103
060103
060103
050104
050104
050105
040105
040105
040105
030105
050106
040106
040106
050107
060107
060107
060108
070108
070108
080108
080109
090109
100110

Follow the procedure shown in Figures 6-23 and 6-24. Formulate the questions; then write the models and all of the cards necessary to answer your questions. Run the problem and answer the question. Remember to use the fictitious Model 90 to answer the question: "Is the RSQ of Model 1 probably zero?"

Plotting Curvilinear Lines

Plotting the lines of best fit that represent a curvilinear relationship is essentially the same as plotting straight lines, which was discussed on pages 150-159. What is done with the squared-term vector, however, needs to be explained. Suppose we find that the following model reflects the relationship between two variables:

$$\tilde{Y}_1 = a_0U + a_1X_1 + a_2X_2 + a_3X_3 + a_4X_4 + a_5X_3^2 + a_6X_4^2$$

where: X_1 and X_2 show group membership; and

X_3 and X_4 contain scores for Groups 1 and 2, respectively.

If we inspect the model, we can see that members of Group 1 will have a value of zero in X_2, X_4, and X_4^2. Then we would have non-zero values in U, X_1, X_3, and X_3^2. We already know that we can pick a value of X_3 as one of our points to graph. If we pick a value of three, the Y value at that point would equal:

$$(a_0*1) + (a_1*1) + (a_3*3) * (a_5*9)$$

The problems on 153-159 involved calculations using the unit vector, a group membership vector, and an interaction vector (the first three terms above). This time the squared-term vector is added and <u>you must remember to square the continuous predictor score</u> before multiplying it by the weight for that vector. Let us look at Problem 6-3 for an example. Model 2 was:

$$\tilde{Y}_1 = a_0U + a_2X_2 + a_3X_2^2$$

There is no group membership variable in this model; so it will be reflected by only one line. The F ratio comparing this model to the full model shows that a cubed relationship actually exists, but we shall plot this model only as an example.

The output shows the weights as follows:

Var. Number	Weight
2	-0.26511654
3	0.05744177
Constant =	5.35613072

Now we need to choose points of X_2 to use for plotting. Note that for a curved line we must use more than two points (which were sufficient for a straight line). Let us use values of 2, 5, 8, and 11 for X_2.

If $X_{2_i} = 2$:

$$\tilde{Y}_{1_i} = \begin{cases} a_0U = 5.36*1 = 5.36 \\ a_2X_{2_i} = -0.27*2 = -0.54 \\ a_3X_{2_i}^2 = 0.06*4 = .24 \end{cases} \quad 5.06$$

If $X_{2_j} = 5$:

$$\tilde{Y}_{1_j} = \begin{cases} a_0U = 5.36*1 = 5.36 \\ a_2X_{2_j} = -0.27*5 = -1.35 \\ a_3X_{2_j}^2 = 0.06*25 = 1.50 \end{cases} \quad 5.50$$

If $X_{2_k} = 8$:

$$\tilde{Y}_{1_k} = \begin{cases} a_0U = 5.36*1 = 5.36 \\ a_2X_{2_k} = -0.27*8 = -2.16 \\ a_3X_{2_k}^2 = 0.06*64 = 3.84 \end{cases} \quad 7.04$$

If $X_{2_1} = 11$:

$$\tilde{Y}_{1_1} = \begin{cases} a_0U = 5.36*1 = 5.36 \\ a_2X_{2_1} = -0.27*11 = -2.97 \\ a_3X_{2_1}^2 = 0.06*121 = 7.26 \end{cases} \quad 9.65$$

Our points on the graph are:

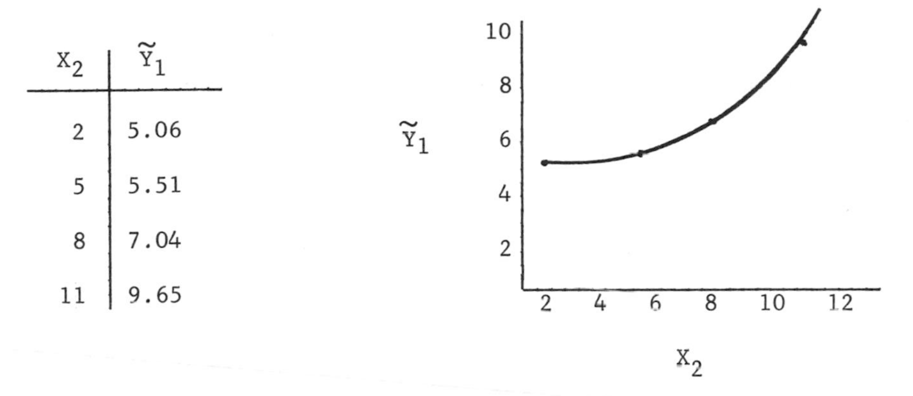

X_2	$\tilde{Y}_1$
2	5.06
5	5.51
8	7.04
11	9.65

Now plot the lines of best fit for the models that represent the state of affairs in Problems 6-3, 6-4, and 6-5. Remember that a cubed-term vector will require that you cube the value chosen before you multiply it by the weight for that vector, just as a squared-term vector requires that you square the value.

<u>Cautions Regarding Testing Sequence</u>

The testing sequence you have just finished was designed to answer a particular set of questions derived from psychological theory. In the particular problem not all possible restrictions were cast. Regarding the full model represented in both Figures 6-23 and 6-24 (the curvilinear interacting state), you will note the two rectilinear vectors were present with two vectors whose elements were the square of the rectilinear elements. We could test the hypothesis that the weights associated with the rectilinear vectors are equal to zero, which would then force the lines into positively or negatively accelerated lines in the right upper-hand quadrant (see Figure 6-25). Also see Figure 6-19c.

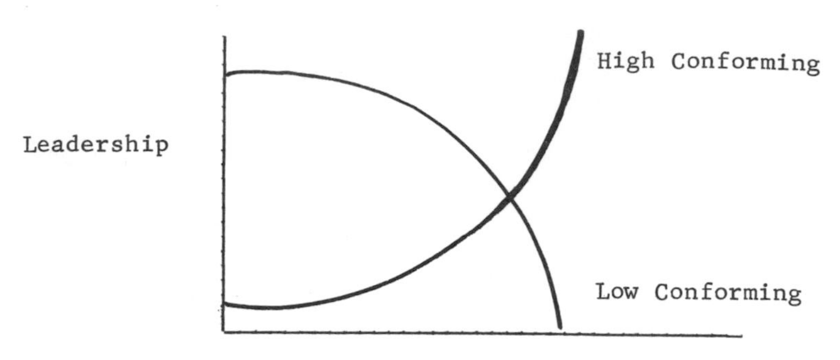

Figure 6-25. A case where the weights associated with the rectilinear vectors are equal to zero but the weights associated with the vectors whose elements are the square of the elements in the rectilinear vectors are not equal to zero and have unlike signs. ($Y_1 = a_0U + a_2X_2 + a_3X_3 + (\pm a_6)X_6 + (\pm a_7)X_7 + E_1$) where:

X_2 = 1 if the corresponding element in Y_1 is from a high conformer; zero otherwise.

X_3 = 1 if the corresponding element in Y_1 is from a low conformer; zero otherwise.

X_6 = the square of the authoritarian score if the corresponding element in Y_1 is from a high conformer; zero otherwise.

X_7 = the square of the authoritarian score if the corresponding element in Y_1 is from a low conformer; zero otherwise.

$a_2 \neq a_3$; a_6 and a_7 have opposite signs.

Essentially, there are a great number of tests one may use for any particular study. The tests one selects should be based upon some theoretical expectations. Indeed, the first step in research, as seen by the present authors, is the generation of a research question. Once the question has been asked, it is a relatively simple matter to cast the several statistical models.

Young researchers we have encountered often come in with data and ask what model should be used or say, "I think a three-way analysis of variance is good to use with these data. What do you think?" Frankly, these people are putting the cart before the horse. We continue to ask, "What is your research question?"

The research question determines (1) the data one collects, and (2) the statistical model one uses.

We hope the present text will minimize the reverse research behavior; but, of course, we have no assurances that individuals will not come up to us and say, "I think my data can use a Figure 6-23 design." We repeat: "What is your research question???"

The Use of Higher Order Polynomials

The vectors whose elements are squared, cubed, quartic, etc., can be used to minimize the errors of prediction. One problem may be anticipated with the use of higher order polynomials:

When the reliability of the original rectilinear vector is moderate (r = .75) or less, the higher order polynomial will geometrically increase the unreliability.

Replication or cross-validation procedures should be employed before interpreting your findings. Conversely, with very highly reliable independent (predictor) variables (r = .98), one would expect the higher order polynomial forms to be replicable.

Measures of height, weight, etc., might represent highly reliable variables whereas measures of attitudes, etc., might give more moderate reliabilities.

Procedures for Cross-Validation

If upon conclusion of a research project you wish to cross-validate your findings, you can get a new sample, collect your data, and generate a vector of predicted scores using the weights of your previous study. Consider the results of Problem 6-2: Test for Interaction. You generated in DATRAN:

```
A(J+6) = A(I+2)*A(I+5)
A(J+7) = A(I+3)*A(I+5)
A(J+8) = A(I+4)*A(I+5)
```

Your full model included three group membership vectors (2, 3, and 4) and three anxiety measures (one for each group 6, 7, and 8). The model accounted for 76 per cent of the variance and the weights were:

$a_2 = .00$

$a_3 = 6.72$

$a_4 = 12.23$

$a_6 = 1.04$

$a_7 = (-.13)$

$a_8 = (-.53)$

constant $a_0 = (-1.36)$

(Note: All values have been rounded to the nearest hundredth. When the variable is a large number, one may wish to go to the third decimal.)

Using the weights given above, we can generate a predicted score vector using these cards in DATRAN: (Remember to start in column 7.)

```
A(J+6) = A(I+2)*A(I+5)
A(J+7) = A(I+3)*A(I+5)
A(J+8) = A(I+4)*A(I+5)
```

```
C = A(I+3)*6.72

D = A(I+4)*12.23

E = A(J+6)*1.04

F = A(J+7)*(-.13)

G = A(J+8)*(-.53)

A(J+9) = C+D+E+F+G-1.36

A(J+10) = A(I+1)-A(J+9)
```

Since a_2 = .000, no multiplication of vector 2 was needed. Look at the vector A(J+9). C, D, E, F, and G summed and added with the constant (-1.36) gives a vector of predicted scores using the weights developed from your previous study. The constants C, D, E, F, and G will not be printed out but are stored and available for the predicted score vector. Vector A(J+9) is a vector of predicted scores since each element in the vector is the sum of the weighted elements (vectors 3 through 8) of the predictor set. If you run this with new data, you do not need model cards because the correlation between the criterion (variable 1) and the predicted score vector (variable 9) when squared will give you your cross-validated R^2. Of course, you can also obtain the R^2 using model cards. If you wish to use the regression equation A(I+1) is your criterion and A(J+9) is your predictor. The R^2 obtained from the initial study and the cross-validated R^2 can give you a fairly direct estimate of shrinkage.

Problem 6-6

In order to get a better understanding of this cross-validation procedure, take the data cards for Problem 6-2 and demonstrate these notions. You already have the first three DATRAN cards given above;

you need the cards C, D . . . A(J+10) = A(I+1)-A(J+9). Punch these seven new cards; change your last two digits of the parameter card from 08 to 10. Submit this new problem with your old data cards (Problem 6-2).

When you get your computer print-out, the correlation between 1 and 9 should be .8717. Square this and compare this value with your print-out from Problem 6-2. Since this is applying the weights upon the same data, the R^2's should be equal within rounding error.

If you use this procedure with a new sample (apply the weights from Problem 6-2 to data from a new sample, find the correlation between the predicted criterion and the actual criterion, and square this correlation for a cross-validated R^2), the cross-validated R^2 will probably be smaller; if it is much smaller, there is some unreliability in the data.

From this print-out, two other interesting observations should be noted:

1. If you square the SD for vector 1, you have total variance (SD_t^2). If you square the SD for vector 9, you have the predictive (among) variance (SD_a^2). Placed in the equation $R^2 = \frac{SD_a^2}{SD_t^2}$, you should get your R^2 = .7596 (a slight difference in the fourth-place due to rounding error).
2. Vector 10 is an error vector. If you square the SD of 10, you have within group variance: $R^2 = \frac{SD_t^2 - SD_W^2}{SD_a^2} = .7596$.

Double Cross-Validation

Replication in behavioral research is difficult to find. In order to encourage replications, we suggest a double cross-validation procedure. If your N is above 400, you might consider randomly splitting the group into two groups, run separate analyses for each group, and then cross-apply the weights using DATRAN as was described above. The squared correlation between the predicted (generated) vector and the criterion gives the cross-validated R^2. If you cross-apply the weights, you get, in essence, two validations and two estimates of the R^2 shrinkage.

For example, we mentioned earlier that higher order polynomials geometrically increase unreliability. When using these more complex models, we can protect ourselves from over-estimating the relationship among predictors and criteria by double cross-validation.

The cross-validated R^2, when compared with the R^2 obtained using the group's own weights, will reveal the stability of the predictive relationship. If more investigators in behavioral research would use this procedure, we might have fewer publications but better quality.

Prediction from only Continuous Variables

Thus far, categorical and/or continuous vectors have been used to predict subjects' criterion scores. What happens when we have no group membership or categorical predictor variables but have a number of continuous predictors? Earlier, relationships between one continuous predictor variable and a criterion variable were investigated. Now let us add another continuous predictor to the model:

$$\tilde{Y}_1 = a_0U + a_1X_1 + a_2X_2 \tag{6.28}$$

where:

$\tilde{Y}_1$ = predicted score on criterion, such as school achievement.

X_1 = anxiety score (continuous).

X_2 = IQ score (continuous).

Depending upon your research interest, you would consider either anxiety or IQ as the range of interest over which you want to predict the criterion scores, and at the same time consider rectilinear influence of the other continuous predictor. If one is primarily interested in investigating anxiety, it would be cast on the X-axis. The graph of the relationship could look like Figure 6-26.

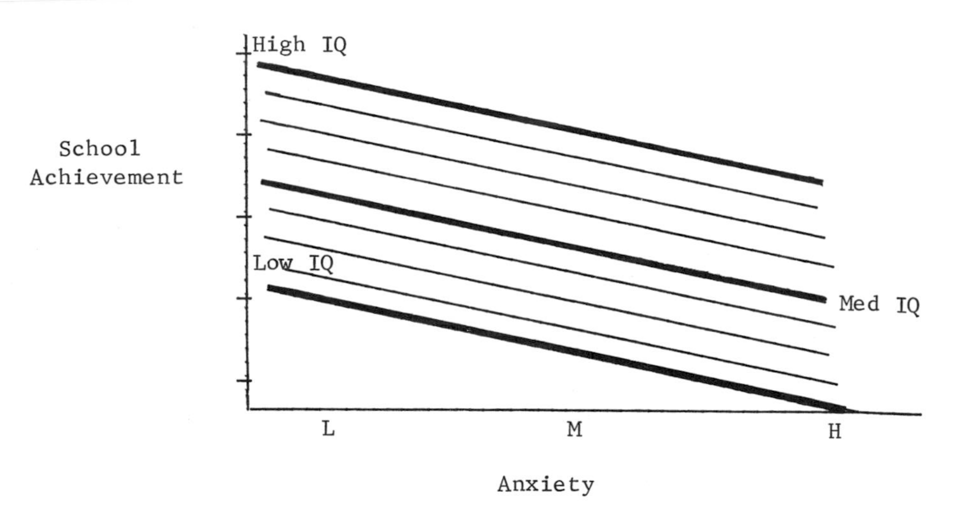

Figure 6-26. A case where there is a constant difference between IQ levels across the observed range of anxiety.

The relationship expressed by Equation 6.28 is represented by a plane surface made up of an infinite number of parallel prediction, or regression, lines. Each line would represent the predicted school achievement scores of persons with a particular IQ score across the range of anxiety scores. You might think of these lines as representing group membership. When we had mutually exclusive categories, we had as many lines as we had groups. Now we have an infinite number of "groups" because the IQ vector is continuous.

To construct such a graph, one might plot only the High, Medium, and Low IQ lines (by choosing IQ values to substitute into the equation for plotting purposes that were representative of High, Medium, and Low (IQ). These are the heavier lines in Figure 6-26.

This is one of the simplest types of relationships which might occur between the three variables. Perhaps there is theoretical or empirical evidence to indicate that the two predictor variables actually interact. Then, we need to cast a model which would allow this interaction to take place. At this time it would be useful to note that no new concepts for constructing models need be introduced here. All of the basic ideas needed to build models containing only continuous variables have already been discussed under combinations of continuous and categorical vectors. The new element is that the models representing continuous variables which are allowed to interact deal with a large number of prediction lines instead of just two or three.

The interaction vector would be formed by multiplying the continuous predictor variables together; the interaction model would look like this:

$$\widetilde{Y}_1 = a_0U + a_1X_1 + a_2X_2 + a_3(X_1*X_2) \qquad (6.29)$$

The graph of this model could be similar to Figure 6-27. As stated previously, either variable can be cast along the X-axis; the prediction equation remains the same. Figure 6-27 shows a graph made by an investigator primarily interested in IQ.

The form of the graph is dependent upon the data. As in Figure 6-26, all the lines are lines of best fit, each for a different anxiety level.

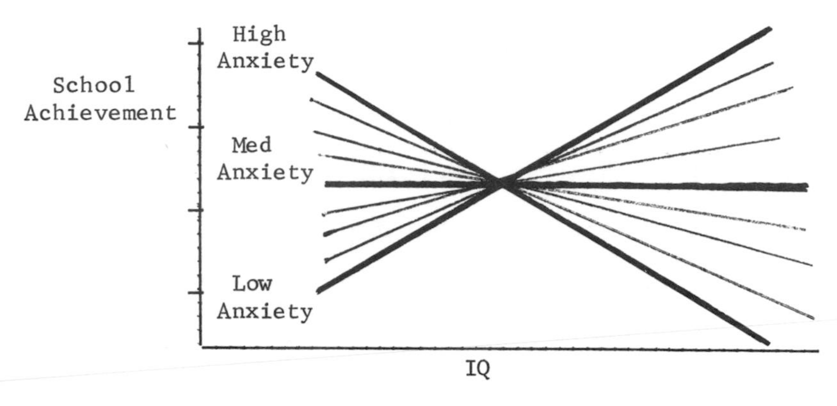

Figure 6-27. A case where there is a non-constant difference between anxiety levels across the observed range of IQ scores.

Earlier in this chapter, anxiety was said to be curvilinearly related to achievement under some conditions. To account for this, we need to build a model to allow the best-fit lines to be curved. The full model, which would allow the interaction of the continuous predictor variables, could be represented by the following:

$$\tilde{Y}_1 = a_0U + a_1X_1 + a_2X_2 + a_3(X_1 * X_2) + a_4{X_1}^2 + a_5(X_2 * {X_1}^2) \quad (6.30)$$

> At this point the predictor equation <u>is</u> affected by the primary interest of the investigator. You will notice that in Equation 6.30, it is the anxiety variable that is squared because anxiety is hypothesized to be curvilinearly related to the criterion. This is the variable of primary interest and will be placed on the X-axis.

To determine if the interaction between anxiety and IQ is significant, Model 6.30 would be compared with a curvilinear model in which the difference between adjacent lines is constant across the range of interest. The model would be:

$$\tilde{Y}_1 = a_0U + a_1X_1 + a_2X_2 + a_3X_1^2 \tag{6.31}$$

A graph of Equation 6.31 could be similar to Figure 6-28.

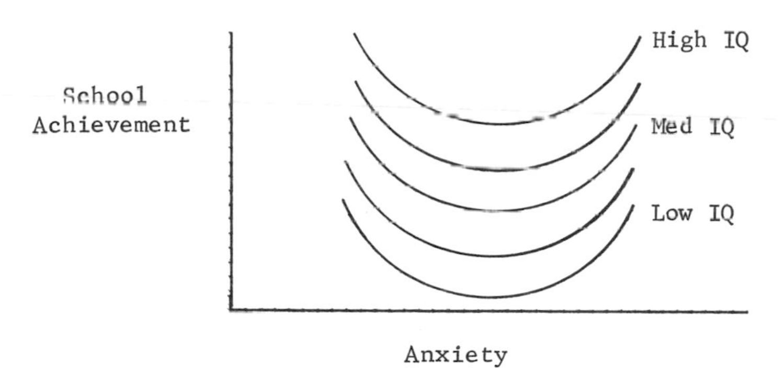

Figure 6-28. Curved parallel lines using continuous IQ data.

As you can see from Equation 6.31, the only weight by which X_2 (IQ) is multiplied is a_2. So, for an increase of one point in the IQ score, the predicted scores would be increased by a value of a_2 across the range of anxiety scores. The lines would, therefore, be parallel. An F test between these two models would indicate if the full model accounts for significantly more variance than the restricted model.

These types of models, or prediction equations, can be expanded to include a number of continuous variables instead of simply two, as shown here. We know that behavior is a function of a number of interacting variables; so the inclusion of at least the more important variables is needed for a better understanding of the specific behaviors in which you are interested. An example of a multivariable problem with continuous variables will now be examined.

Whiteside (1964), in an unpublished doctoral dissertation, attempted to predict high school success from ninth-grade information. Whiteside selected a number of variables which theoretically should predict school success and cast a series of equations to reflect the relationship. The final equation which cross-validated was:

$$\tilde{Y}_1 = -2.574 + .028(X_1) + .120(X_2) + .019(X_3) + .011(X_4) + .011(X_5)$$

where:

$\tilde{Y}_1$ = predicted high school GPA;

X_1 = score on mutilated word test;

X_2 = score on nomination: academic model;

X_3 = 1 is subject were a girl; 0 if a boy;

X_4 = CTMM score times STEP Listening score divided by 100;

X_5 = ninth-grade GPA squared.

This model was derived from information about students in four different communities. It accounts for about 70 per cent of the variance on high school GPA. A practical use for it could be selection of predicted failures so that some type of special attention could be given them in hopes of upsetting the prediction.

From this model we see that, for his sample, ninth-grade GPA is curvilinearly related to high school GPA. The California Test of Mental Maturity (CTMM) interacts with STEP Listening scores to account for part of the variance. He also found that the sex of the child accounts for some of the GPA variance unaccounted for by the other variables. Rectilinearly related to high school GPA are scores on mutilated word test, nomination: academic model, and sex.

Multivariable Use of Categorical Predictors

Often in research involved with human behavior, the investigator encounters data that emerge in categorical form: Example -- (1) Sex: Male, Female; (2) Treatment: A, B, C, etc.; (3) Education: Grammar school, junior high, high school, college; (4) Health: Good, bad; (5) School Classification: Drop-out, graduate; (6) Crime: Delinquent, not delinquent, etc.

Furthermore, individuals trained in traditional analysis of variance frequently take continuous data (e.g., anxiety measures) and classify the scores into two or more mutually exclusive groups (e.g., high-anxiety, moderate-anxiety, low-anxiety).

In Chapter IV we used categorical data utilizing one variable, which was treatment condition. In Problem 6-2, we had three treatment groups (categorical data) and a continuous vector representing anxiety scores as predictors of performance on a test. If we break the continuous anxiety vector into three categories, we have an example of multivariable use of categorical predictors.

> The research question might be: "Is the influence of treatments non-constant and different across the three levels of anxiety?"

Figure 6-29 shows a case where the influence of treatment is non-constant and different across the three levels of anxiety. Look at Figure 6-15. Note the similarity between the two figures.

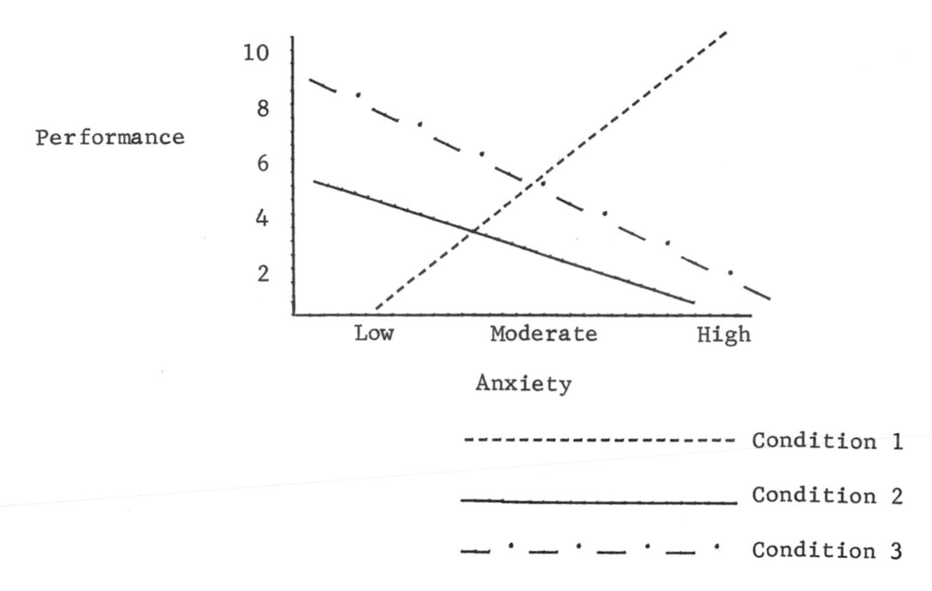

Figure 6-29. A possible finding using a two-variable analysis with categorical data.

If we have three treatments and three levels of anxiety, in order to reflect the points in Figure 6-29 we need nine categorical vectors. The boxes below show this:

	A_1	A_2	A_3
T_1			
T_2			
T_3			

$$T_1A_1 + T_1A_2 + T_1A_3 + T_2A_1 \ . \ . \ . \ T_3A_3$$

Equation 6.32 shows the full model used to predict the nine points:

$$(6.32) \qquad Y_1 = a_0U + a_1X_1 + a_2X_2 + a_3X_3 + a_4X_4 \ . \ . \ . + a_9X_9 + E_1$$

where:

X_1 = 1 if the corresponding score on Y_1 comes from an individual T_1A_1; zero otherwise.

X_2 = 1 if the corresponding score on Y_1 comes from an individual T_1A_2; zero otherwise.

X_3 = 1 if the corresponding score on Y_1 comes from an individual T_1A_3; zero otherwise.

X_4 = 1 if the corresponding score on Y_1 comes from an individual T_2A_1; zero otherwise.

.

.

.

X_9 = 1 if the corresponding score on Y_1 comes from an individual T_3A_3; zero otherwise.

$E_1 = (Y_1 - \widetilde{Y}_1)$.

To answer the question: "Is the influence of treatments non-constant and different across the three levels of anxiety?", we must force the difference between points to be constant:

$$T_1A_1 - T_2A_1 = T_1A_2 - T_2A_2 = T_1A_3 - T_2A_3 \quad \text{and}$$
$$T_1A_1 - T_3A_1 = T_1A_2 - T_3A_2 = T_1A_3 - T_3A_3$$

Figure 6-30 shows the possible state of affairs.

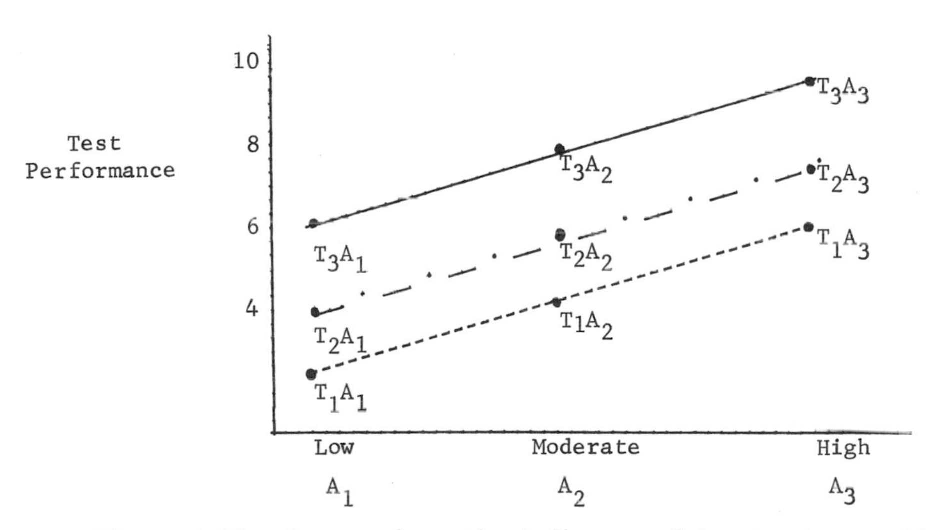

Figure 6-30. A case where the influence of treatments provides a constant difference across the three levels of anxiety.

In order to reflect Figure 6-30, all subjects who are low-anxious must share a common weight and likewise all moderately-anxious subjects and all high-anxious subjects. This can be reflected in Equation 6.33.

(6.33) $$Y_1 = a_0U + a_1T_1 + a_2T_2 + a_3T_3 + a_4A_1 + a_5A_2 + a_6A_3 + E_2$$

where: T_1, T_2, and T_3 represent categorical vectors for treatment, and

A_1, A_2, and A_3 represent categorical vectors for levels of anxiety.

If the F test between the models expressed in Equations 6.32 and 6.33 is significant, the restricted model significantly increases errors of prediction; therefore, an interaction state of affairs exists.

A plotting of the nine points will show the relationship. The relation may take the form of Figure 6-29 or 6-31. Note in Figure 6-29 we have a rectilinear interaction and in 6-31 a possible curvilinear interaction. This illustrates the fact that when the full model uses categorical data, no assumption of rectilinearity is made since each point projected has a unique weight (nine weights, one for each of the nine points). If you refer back to Chapter IV, you will note that these points represent the respective cell means.

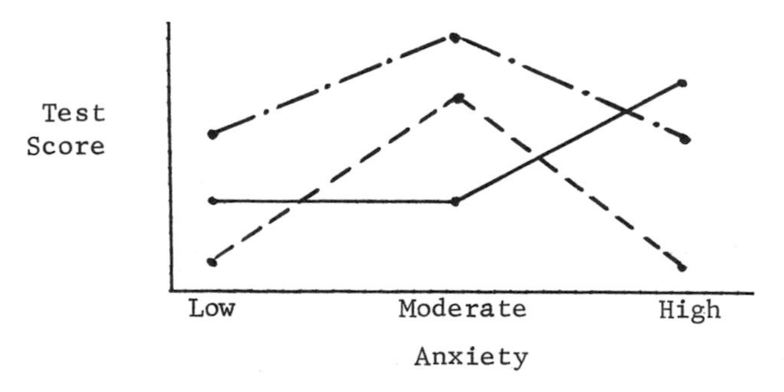

Figure 6-31. A possible finding when one rejects the constant difference hypothesis.

Program LINEAR and the Multivariable Design

Let us assume you have the data punched on a card in the following format: (2X,6F1.0,F2.0)

where:

col 1-2 = identification number.

variable X_1 3 = 1 if Treatment 1; zero otherwise.

X_2 4 = 1 if Treatment 2; zero otherwise.

X_3 5 = 1 if Treatment 3; zero otherwise.

X_4 6 = 1 if low-anxious; zero otherwise.

X_5 7 = 1 if moderate-anxious; zero otherwise.

X_6 8 = 1 if high-anxious; zero otherwise.

X_7 9-10 = score on test.

Equation 6.33 can be cast into the following regression equation:

$$(6.34) \qquad Y_1 = a_0U + a_1X_1 + a_2X_2 + a_3X_3 + a_4X_4 + a_5X_5 + a_6X_6 + E_2$$

where: Y_1 = variable 7.

All low-anxious subjects will share the same weight regardless of treatment. See the several vectors below:

$$\begin{array}{c} \text{Subject} \\ 1\\2\\3\\4\\5\\6\\7\\8\\9 \end{array}
\quad
\overset{Y_1}{\begin{bmatrix} \\ \\ \\ \\ \\ \\ \\ \\ \\ \end{bmatrix}}
= a_0 \overset{U}{\begin{bmatrix}1\\1\\1\\1\\1\\1\\1\\1\\1\end{bmatrix}}
+ a_1 \overset{X_1}{\begin{bmatrix}1\\1\\1\\0\\0\\0\\0\\0\\0\end{bmatrix}}
+ a_2 \overset{X_2}{\begin{bmatrix}0\\0\\0\\1\\1\\1\\0\\0\\0\end{bmatrix}}
+ a_3 \overset{X_3}{\begin{bmatrix}0\\0\\0\\0\\0\\0\\1\\1\\1\end{bmatrix}}
+ a_4 \overset{X_4}{\begin{bmatrix}1\\0\\0\\1\\0\\0\\1\\0\\0\end{bmatrix}}
+ a_5 \overset{X_5}{\begin{bmatrix}0\\1\\0\\0\\1\\0\\0\\1\\0\end{bmatrix}}
+ a_6 \overset{X_6}{\begin{bmatrix}0\\0\\1\\0\\0\\1\\0\\0\\1\end{bmatrix}}
+ \overset{E_2}{\begin{bmatrix} \\ \\ \\ \\ \\ \\ \\ \\ \\ \end{bmatrix}}$$

Note that Subject 1 has two weights in addition to the regression weight (a_0). Subject 1 is in Treatment 1 ($\underline{X_1}$) and is low-anxious ($\underline{X_4}$). Subject 4 is in Treatment 2 ($\underline{X_2}$) and is low-anxious ($\underline{X_4}$). Subject 7 is in Treatment 3 ($\underline{X}_3$) and is also low-anxious; therefore, there will be three points for the low-anxious level (one for each treatment) and the distance between these points will be determined by the weights (a_1, a_2, and a_3) associated with treatment. The same will hold for the other two anxiety levels; thus, we have equal distances between the groups across the levels of anxiety. The weights associated with anxiety (a_4, a_5, and a_6) will determine the configuration of the parallel lines drawn between the points. Just as in the case of interaction using multivariable categorical vectors, the constant difference does not assume rectilinearity. Figure 6-30 may be appropriate, but Figure 6-32 might be the state of affiars. We have to plot the points to determine this.

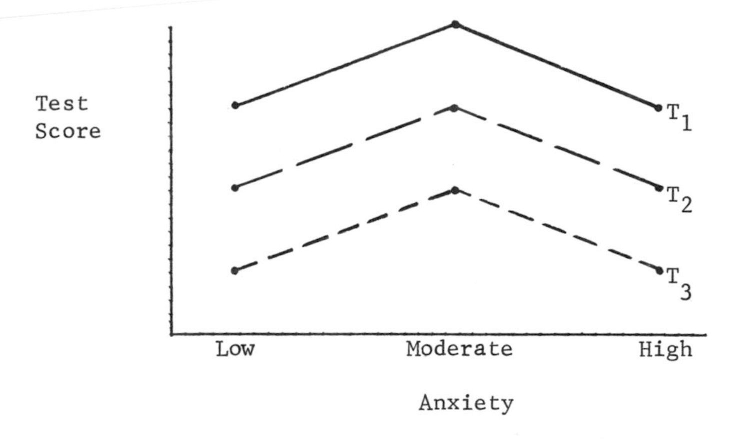

Figure 6-32. A case where the constant difference case is non-rectilinear.

Figure 6-32 could be a case of parallel curved lines (if the anxiety vector was continuous).

In order to form the nine vectors representing the interaction model (Equation 6.32), we can generate in DATRAN the nine new vectors. If we multiply the elements in vector 1 times the elements in vectors

4, 5, and 6; vector 2 times 4, 5, and 6; and vector 3 times 4, 5, and 6, we get nine mutually exclusive categories.

The DATRAN statements take the form:

```
A(J+8) = A(I+1)*A(I+4)
A(J+9) = A(I+1)*A(I+5)
A(J+10) = A(I+1)*A(I+6)
A(J+11) = A(I+2)*A(I+4)
.
.
.
A(J+16) = A(I+3)*A(I+6)
```

Model One, the full model, will be cast:

Model 1   003001.E-05081607   (Equation 6.32)

and Model Two, the restricted:

Model 2   003001.E-05010607   (Equation 6.33)

The degrees of freedom associated with the numerator = 9-5 = 4. (The full model has one linear dependency with the vector U. The restricted model has two linear dependencies with U -- one for anxiety and one for treatment. You might wish to return to Chapter III to demonstrate this.)

Of course, if the constant difference restriction is accepted, you might wish to determine if the constant difference equals zero. You should be able to anticipate the restriction $a_1 = a_2 = a_3 = a_0$. (All treatments share a common weight.)

An easy way to cast your models in multivariable categorical form is to first graph out in box form as we did the T*A illustration. Count the cells and you have the number of vectors needed to express your unrestricted model. These analyses are similar to traditional analysis of variance, with the exception that when the interaction is not significant, we pool that term into the error. (If you have not been introduced to traditional analysis of variance, disregard the previous comment.)

The Case Where Multivariable Forms Are Inappropriate

The treatment by anxiety study just used had one dimension (anxiety) which was ordinal. That is, we had high, moderate, and low. Under these circumstances, the interaction and curved forms can provide psychological meaning. On the other hand, if we have two nominal dimensions (no rational order), we should use a univariate analysis. For example, if one has three different treatments with no obvious order and two sexes, a box of the form given below is meaningless:

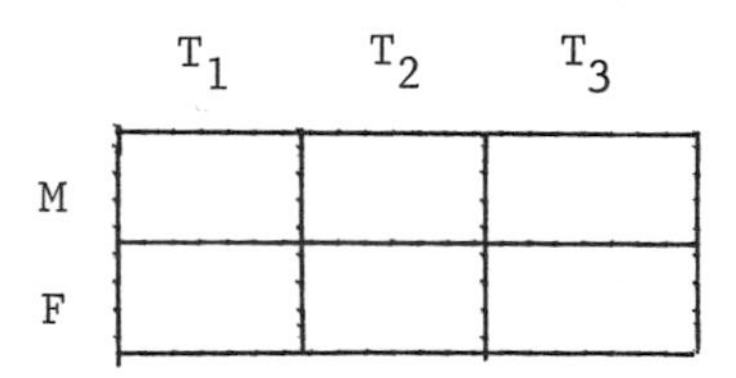

A graph like Figure 6-33 is also meaningless since the order of treatment is arbitrary. We do not intend to suggest that no ordinality can be attributed to treatment; indeed, if you have increasing amounts of training across treatment, then some ordinality can theoretically give Figure 6-33A meaning.

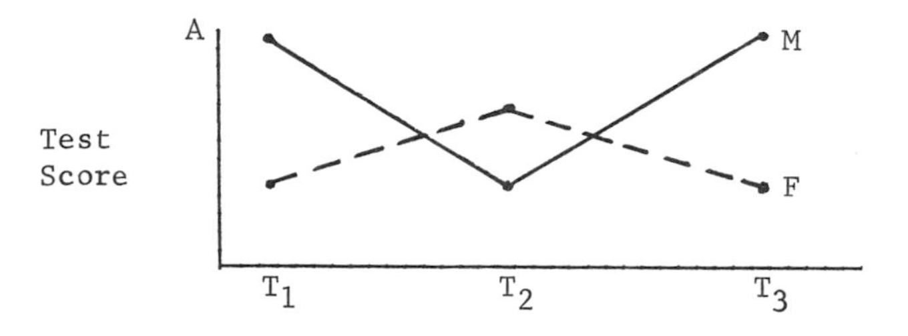

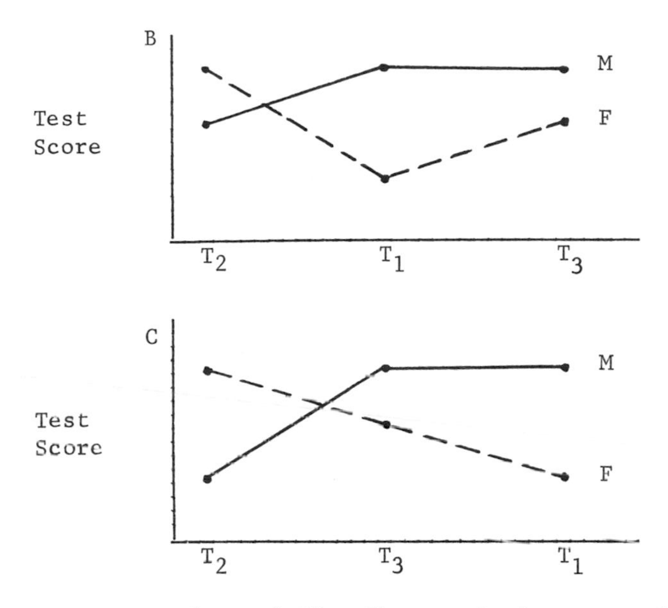

Figure 6-33. The meaninglessness of plotting the ordinal data into a multivariable form.

Nevertheless, when both are nominal dimensions, you can generate the appropriate vectors to represent each cell (in the case above -- 6 vectors) and hypothesize no difference either by sex or treatment.

To illustrate:

$$Y_1 = a_0U + a_1X_1 + a_2X_2 + a_3X_3 + a_4X_4 + a_5X_5 + a_6X_6 + E_1$$

where:

$X_1 = T_1$ female (1; zero otherwise).

$X_2 = T_2$ female (1; zero otherwise).

$X_3 = T_3$ female (1; zero otherwise).

$X_4 = T_1$ male (1; zero otherwise).

$X_5 = T_2$ male (1; zero otherwise).

$X_6 = T_3$ male (1; zero otherwise).

Y_1 = some criterion.

One might ask: "Is sex unrelated to treatments?" Then a_1 = a_4; a_2 = a_5; and a_3 = a_6. Using DATRAN you can add X_1 and X_4, X_2 and X_5, and X_3 and X_6. These three vectors represent the treatments. If a significant amount of information is lost, then some treatment effects are influenced by sex in some manner. One may wish to explore all possibilities; but remember, if you cannot defend ordinality in the variables from your theory, look upon the analysis in a univariate form as was presented in Chapter IV.

Review

Chapters V and VI include a number of mechnaical procedures which take some time to master. You can overcome most of the mechanical problems by referring back to these two chapters. The most difficult aspect of using the generalized multiple linear regression approach or any research approach relates to the formation of research questions and the casting of statistical models to reflect the question. In that the research question depends upon a theory of behavior, a text of this type can provide hints regarding question formation. Your knowledge of your field must provide the relevant variables and the expected functional relationships.

In order to illustrate the sequence of events in research, two examples are provided below. We will first discuss the study in terms of the questions asked. Then we will examine the mechanical process by which the questions were answered.

I. Comparative Advantages of Unlike Exercises in Relation to Prior Individual Strength Level (Bender, Kelly, Pierson, and Kaplan, 1968)

Discussion of the Question

Bender, et al., were confronted with the problem that non-contact, serious knee injuries at West Point were highest among cadets who were in the lowest quartile in strength in a knee extension test. Furthermore, the majority of the injuries occurred during the first few months of training. Exercise designed to strengthen the knee tended to reduce injury. The problem was to find an exercise program which would provide the most efficient increase in strength. The literature indicated that in some cases isometric exercises were more effective in strength improvement than isotonic (i.e., stool stepping), and other research studies found the opposite to be the case. Speculation in the literature suggested that the contradictory findings might be due to individual differences. In view of the empirical data and theoretical speculation, it was evident not only that the research program should investigate treatment effects but also that some control for pre-treatment strength level should be used. Traditional research in the field often controlled for pre-test differences by looking at gain scores through the use of a correlated "t." If one treatment is best for the weak cadet and the other treatment is best for the strong cadet, the correlated "t" would not detect these effects. In essence, the theory would imply that the two treatments (isotonic and isometric) might not have a common regression slope across the observed pre-test strength levels. Furthermore, the two treatments might not have a common Y-intercept.

A semi-formal statement of the problem was:

> "This study was designed to determine the effectiveness of isometric contraction and isotonic movement for strength development as related to the strength level of the individual prior to application of the exercise regimens."

A formal statement of the problem was:

> "Is there a constant difference in the effects of the two exercise treatments on the second testing period, across the range of observed strength level at time one?"

In lay terms, the formal question asks: "If one procedure improves strength better for the weak cadet, does it also yield the same increase for the strong cadet?" Figure 6-34a represents the question. If $d_1 = d_2$, then the co-linear (parallel-line) case can be accepted. Of course, to answer the question, the non-colinear (interaction) equation must be cast to compare with an equation which forces co-linearity.

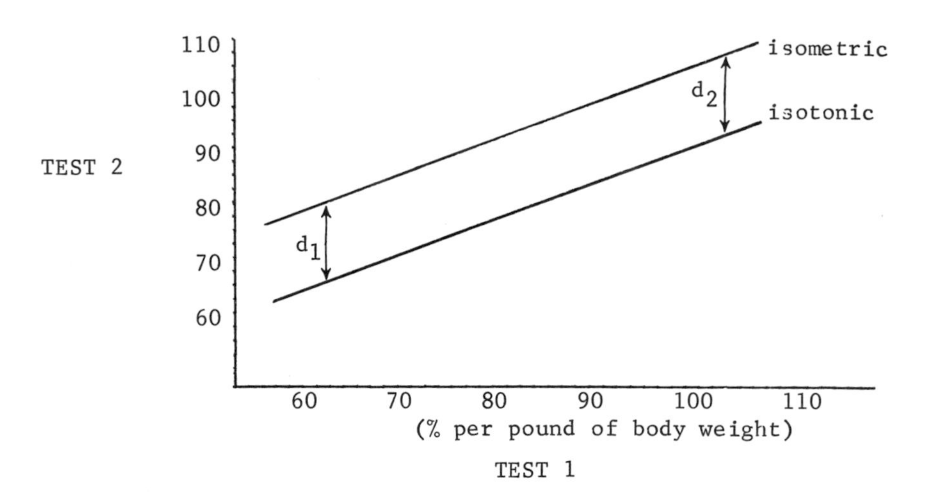

Figure 6-34a. A co-linear, or constant, difference case (knee extension at 90° position).

Earlier in this chapter we showed that two non-parallel lines can be cast from one equation. In this case we have two groups: (1) isometric and (2) isotonic (the original study included a control group, but for expository purposes we shall not include it). We also have a measure of pre-training strength. If we reflect the co-linear model expressed in Figure 6-34a, we get:

(6.35) $$\tilde{Y}_1 = a_0U + a_1T_1 + a_2T_2 + a_3X_1$$

where:

T_1 = 1 if the score on Y_1 comes from a member of the isometric treatment group; zero otherwise.

T_2 = 1 if the score on Y_1 comes from a member of the isotonic treatment group; zero otherwise.

X_1 = pre-test weight-strength ratio.

$\tilde{Y}_1$ = post-test weight-strength ratio, predicted.

Since the two groups share a common weight (a_3) associated with X_1, the slopes by necessity must be the same. Can you construct an equation to allow each treatment to have its own slope? We shall assume you can do this. At the end of this section the model is given as Model One (1).

Figure 6-34b shows the state of affairs found in the Bender study. The regression lines were not co-linear. An inspection of Figure 6-34b shows that if the cadet was below the mean on the pre-test, stool-stepping produces the greatest final strength. Conversely, if the cadet was initially above average, the isometric contraction produced greater final strength.

Consequently, the recommendation to West Point officials was: The strength training program to reduce injury might best be (1) stool-stepping for cadets below average in strength; and (2) isometric contraction for the cadets who are initially above average in strength.

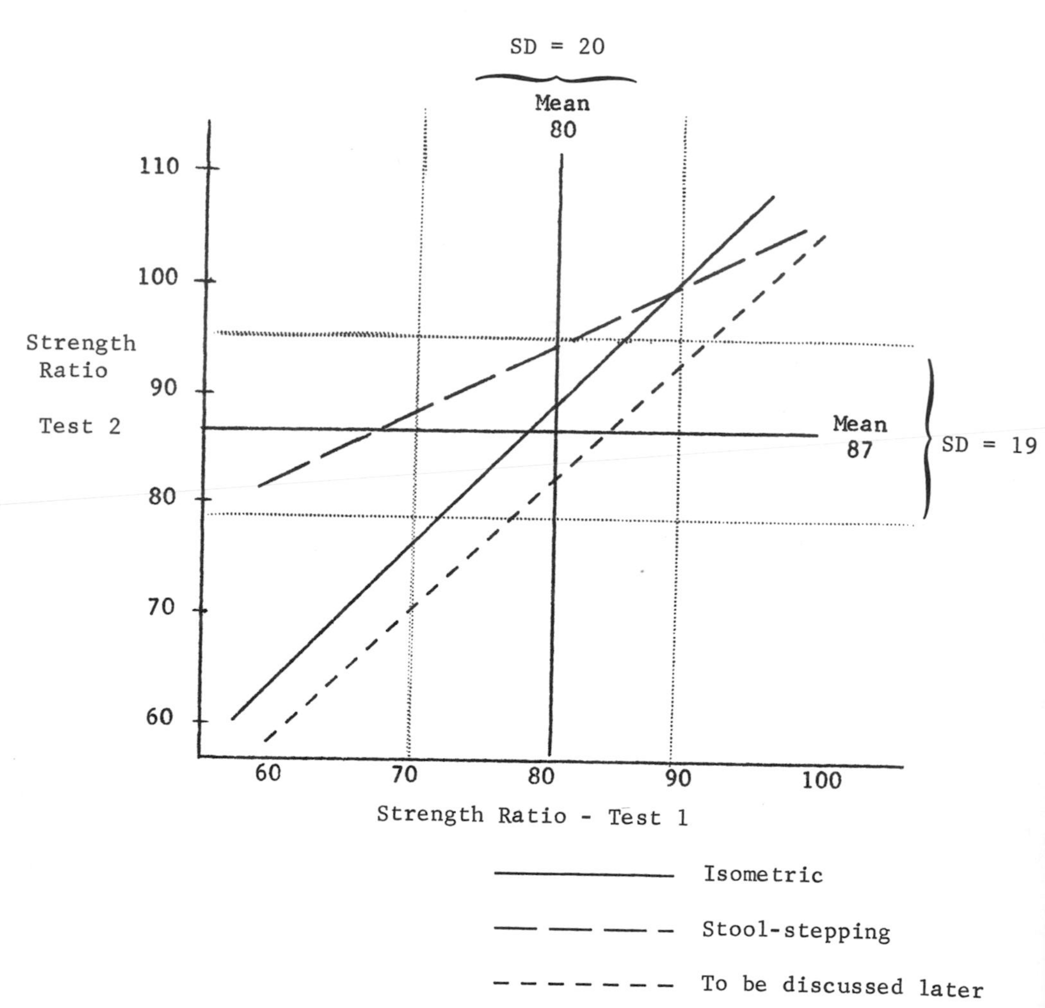

Figure 6-34b. The condition found in the Bender, et al., study (knee extension at 90° position -- Test 1 and Test 2).

Process of Answering the Question

The research question was derived from the available evidence and theory. The relevant data were collected and a full model was constructed to fit a set of lines which Bender, _et al._, expected from the theory as reflected in the research question. The full model was restricted to eliminate the interaction of treatment with initial strength levels. The restriction was rejected since a significant loss of information was found.

Model 1 $\tilde{Y}_1 = a_0U + a_1T_1 + a_2T_2 + a_3T_1{*}X_1 + a_4T_2{*}X_1$

Co-linearity was tested by imposing the restriction $a_3 = a_4 = a_5$, a common weight.

$$\tilde{Y}_1 = a_0U + a_1T_1 + a_2T_2 + a_5T_1{*}X_1 + a_5T_2{*}X_1$$

but since $T_1{*}X_1 + T_2{*}X_1 = X_1$ and a common weight is imposed, this can be expressed:

Model 2 $\tilde{Y}_1 = a_0U + a_1T_1 + a_2T_2 + a_5X_1$

In the original Bender study three groups, with 65 cadets in each, were used. If the co-linear model $(\tilde{Y}_1)$ in Figure 6-35 had been tenable, then a second restricted model would have been used to answer the question: "Is the constant treatment difference equal to zero?"

Model 3 $\tilde{Y}_1 = a_0U + a_7X_7$

This model was run in the computer at the same time as the full and first restricted models. The investigators could then examine it if the first restriction had been tenable.

Figure 6-35 shows the three models and the R^2 value which led to the drawing of Figure 6-34b. The results of the comparison between Model 2 and Model 3 are not reported since the restriction $a_4 = a_5 = a_6 = a_7$, a common weight, was _not_ tenable.

Model	R^2	df	F	P
$\tilde{Y}_f = a_0U + a_1T_1 + a_2T_2 + a_3T_3 + a_4X_4 + a_5X_5 + a_6X_6$	.81			
$\tilde{Y}_{r_1} = a_0U + a_1T_1 + a_2T_2 + a_3T_3 + a_7X_7$	.79	2/170	8.87	.01
$\tilde{Y}_{r_2} = a_0U + a_7X_7$	not appropriate			

T_1 = 1 if score on Y comes from a member of isometric treatment; zero otherwise.

T_2 = 1 if score on Y comes from a member of isotonic treatment; zero otherwise.

T_3 = 1 if score on Y comes from a member of control treatment; zero otherwise.

X_4 = score on pre-test if the score on Y comes from a member of the isometric group; zero otherwise.

X_5 = score on pre-test if the score on Y comes from a member of the isotonic group; zero otherwise.

X_6 = score on pre-test if the score on Y comes from a member of the control group; zero otherwise.

$X_7 = X_4 + X_5 + X_6$.

Figure 6-35. Models, F-tests, and R^2 for the two questions regarding knee extension - 90°.

Mechanically, all three models were cast in program LINEAR and the two F tests were made. This, of course, saves time. When the restriction increased the errors of prediction ($p < .01$), then the weights of the larger model (in this case, Model 1) were used to cast the lines of best fit.

II. Behavioral Problems and Change in Verbal IQ
(Phillips, Adams, and McNeil, 1967)

You may recall in our previous discussion that many investigators make a very drastic but common mistake when they force a rectilinear relationship upon their data. One of the most common instances of this occurrence involves the use of the Pearson Product correlation coefficient. Whenever a researcher finds a low Pearson correlation coefficient, the conclusion of no relationship is drawn. The correct conclusion is that there is no rectilinear relationship between these variables for these subjects. As you might suspect, there could very well be a curvilinear relationship, even though a rectilinear relationship does not exist. The data may even yield a significant curvilinear relationship over and above the linear relationship. That is, there may be a curvilinear component as well as a linear component.

Phillips, Adams, and McNeil (1967) were interested in the effects on school-related variables as a function of change in verbal intelligence. The amount of variance that the investigators were able to account for with the rectilinear relationship was discouragingly low. As a result of some theoretical and statistical considerations, the investigators decided to consider the second-degree relationships between the predictor variable and the dependent variable. The theoretical considerations involved to some extent the idea that children exhibiting (or admitting to) behavioral problems will not care about performing well on an intelligence test, nor will they be able to make the normal gains expected of grade school children. Conversely, the children who do not present behavioral problems in school are generally the more capable children who are performing at their peak levels at all times. These two groups were expected to change little in verbal intelligence over a two-year period. The large increases in verbal intelligence were expected of those children who present a moderate amount of behavioral problems. These children are willing to work hard to succeed in school, and they, in fact, have to work hard to do so. For these reasons, the investigators hypothesized that these would be the children who would gain more in verbal intelligence over the two-year period. The statistical considerations involved the concept of regression effects in change scores (see Harris, 1963), and the desire to account for more variance in the criterion variable. Figure 6-36 presents a schematic description of the hypothesis.

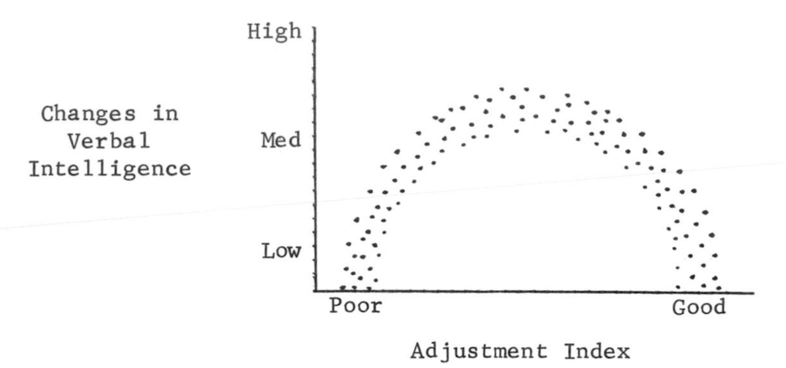

Figure 6-36. A scatter plot which reflects one possible second-degree relationship.

The full regression model should allow for the second-degree relationship, as well as the linear relationship:

$$\tilde{Y}_1 = a_0U + a_1X_1 + a_2X_1^2$$ (Full model for second-degree relationship)

where:

X_1 = the change in verbal intelligence.

$\tilde{Y}_1$ = the predicted behavioral index.

To determine if there is a significant second-degree relationship (over and above the linear relationship), the following restriction must be made: $a_2 = 0$. Imposing this restriction on the above full model, we arrive at the following restricted model:

$$\tilde{Y}_1 = a_0U + a_1X_1$$ (Restricted model which reflects rectilinear relationship)

We note again that the above model yields the simple Pearson Product-Moment correlation. The results of applying these two regression models to the data obtained from 107 Latin American sixth-graders from Austin, Texas, is presented in Figure 6-37.

	Rectilinear Correlation	Correlation Plus Rectilinear Correlation
Grade Point Average	.18	
School Motivation	.17	
MAT Non-Verbal Achievement	*.23	
MAT Verbal Achievement	*.35	
School Anxiety	-.08	
Self-Disparagement in Relation to Peers	.03	*.19
Avoidance Style of Defensiveness	-.12	
Approach Style of Defensiveness	.06	
Feelings of Inferiority	*.28	
Neurotic Symptoms, Academic	*.28	*.34
Neurotic Symptoms, Social	*.34	*.49
Aggression with Independent Strivings	.15	
Active Withdrawal	*.21	*.34
Emotional Disturbance with Depression	.08	
Self-Enhancement through Derogation of Others	.04	
Diffuse Hyperactivity	.13	*.32
Peer Acceptance	.07	*.25
Peer Rejection	.07	

*Indicate correlation coefficient significantly different from 0 at .05 level of confidence.

Figure 6-37. School behavioral correlates of change in verbal intelligence.

The important aspect of Figure 6-37 is that in some cases we are accounting for much more variance with the full model than with the restricted model. We should also note that in several cases there is a significant curvilinear relationship, as well as a linear relationship.

With respect to three variables (avoidance-style of defensiveness, diffuse hyperactivity, and peer acceptance), the inference of no relationship from the rectilinear model does not agree with the inference of a statistical relationship from the curvilinear model.

The reader should note that only one form of curvilinearity was investigated in this study -- that of a second-degree nature. This type of curvilinearity corresponded to the theoretical and statistical considerations and was, thus, the relationship that was investigated. Other theoretical considerations may lead to third, fourth, and higher order curvilinear relationships.

Chapter VII Covariance Statistical Control and Multivariate Techniques

Covariance

There are two general ways in which a researcher can control for experimental error. One technique for controlling experimental error is a direct control, sometimes referred to as "experimental control." In this technique, the researcher considers all relevant or concommitant variables in designing his study. Any variable of importance to the researcher is taken into consideration, or controlled for, before the data is collected. This requires that the researcher have complete control as to who his subjects are (in regard to the relevant variables) and the manner in which they will participate in the study. In many instances, however, the researcher is unable to control his selection of subjects to the extent that he desires. When this occurs, the researcher can use the second method of controlling experimental error due to selected variables. He may use a form of statistical control for the selected variables that were not controlled for in the experimental design. This statistical control is facilitated by making appropriate adjustments to the criterion means for the various treatment conditions. This method of controlling experimental error is called analysis of covariance. This method will be discussed first in terms of conventional analysis of variance and then in terms of multiple linear regression analysis.

As an example of the need for statistical control in some experiments, we will make use of the data presented in Chapter VI (page 202). In this example, you may recall, the research question was suggested to be: "Is the influence of treatments non-constant and different across the three levels of anxiety?" To investigate this question, the researcher designed his study such that the cell frequencies were equal or proportional from treatment to treatment. (Note: The researcher did this to satisfy the assumptions required in the traditional analysis of variance model.) The researcher had the opportunity to manipulate his subjects according to his experimental design. While he was in the process of collecting his data, the researcher realized that age should have been taken into consideration as a concommitant variable.

In order to statistically control the concommitant variable, the researcher was required to compute: (1) the correlation between the concommitant variable (age) and the criterion variable (score on the test); and (2) the coefficient of regression of the criterion on the concommitant variable. Given this information, the researcher could adjust the criterion measures for the effects of the concommitant variable. (A complete discussion of the techniques for determining the adjusted criterion scores may be found in Winer, 1962.) After the researcher adjusted the criterion scores, the traditional analysis of variance techniques were applied to the "adjusted" criterion scores.

To statistically control the concommitant variables (age), the researcher must take into consideration the following assumptions (Lindquist, 1956):

1. The subjects in each treatment group were originally either (a) selected at random from the same parent population; or (b) selected from the same parent population on the basis of their concommitant measures only.
2. The concommitant measures are unaffected by the treatments.
3. The criterion measures for each treatment group are a random sample from those for a corresponding treatment population.
4. The regression of the criterion measure on the concommitant measure is the same for all treatment populations.
5. The regression in assumption #4 is linear.
6. The distribution of adjusted scores for each treatment population is normal.
7. The distribution of adjusted scores have the same variance.

These assumptions for analysis of covariance are much more rigorous than the assumptions underlying the analysis of variance model given in Chapter IV. The first assumption concerning the manner in which the treatment groups are selected has been violated often and the violation has led to false conclusions. If the treatment groups are selected at random, the adjustments made in the criterion scores are made only for _chance_ differences in the control variable. All too often researchers assume that the adjustment is made for systematic differences existing among the treatments with respect to the concommitant variable before the treatments are administered. This systematic difference is included in the adjustment only when the treatment groups are selected with respect to the concommitant variable. In the example being discussed, the researcher did not select the treatment groups with respect to the concommitant variable because the concommitant variable (age) was not considered until after the collection of data had begun. As indicated before, the only difference accounted for in the adjusted scores would be the chance difference in the concommitant variable. Usually in an experiment of this kind, the statistical control desired would be the systematic differences among the treatment groups with respect to the concommitant variable.

Covariance Handled by LINEAR

Usually analysis of covariance is a difficult topic for the beginning student. The concepts of "adjusted" scores, chance differences, and systematic differences are difficult to comprehend. In addition, the beginning student must be able to handle the analysis of variance techniques in order that he might correctly deal with the "adjusted" criterion scores.

In developing materials for this text, the authors have found that the covariance analysis technique is more easily understood when the discussion deals with concommitant variables and the proportion of variance accounted for by these variables. The discussion of covariance is essentially a discussion of multivariable use of categorical predictors. The following discussion will deal with the same problem discussed earlier in the chapter.

If the researcher is concerned that the concommitant variable (age) may influence his results, he is actually expressing concern as to whether or not he can account for a large proportion of variance in the criterion vector with only the knowledge of the age of the individuals. Another way to express the concern of the researcher is to ask: "Does the additional knowledge of the treatment group and anxiety level of the individual significantly improve the prediction of the criterion variable (test score) over and above that which the concommitant variable (age) predicts?" This question can easily be answered with the following sequence of models where age is always included in the equation as a covariate:

Full model:

$$Y_1 = a_0U + a_1X_1 + a_2X_2 + a_3X_3 + a_4X_4 + a_5X_5 \ . \ . \ . + a_{10}X_{10} + E_1$$

where:

X_1 = age of the individual.

X_2 = 1 if the corresponding score on Y_1 comes from an individual T_1A_1; zero otherwise.

X_3 = 1 if the corresponding score on Y_1 comes from an individual T_1A_2; zero otherwise.

X_4 = 1 if the corresponding score on Y_1 comes from an individual T_1A_3; zero otherwise.

X_5 = 1 if the corresponding score on Y_1 comes from an individual T_2A_1; zero otherwise.

.

.

.

X_{10} = 1 if the corresponding score on Y_1 comes from an individual T_3A_3; zero otherwise.

E_1 = $(Y_1 - \tilde{Y}_1)$.

To answer the question: "Is the influence of treatments non-constant and different across the three levels of anxiety in the presence of age?", we have the following restricted model:

$$Y_1 = a_0U + a_1X_1 + a_2T_1 + a_3T_2 + a_4T_3 + a_5A_1 + a_6A_2 + a_7A_3 + E_2$$

where:

X_1 = the age of the individual.

T_1, T_2, and T_3 represent categorical vectors for treatment.

A_1, A_2, and A_3 represent categorical vectors for levels of anxiety.

If the F test is significant, interaction exists.

If the F test is not significant, then the question to be asked may be stated: "Is the influence of treatments different across the three levels of anxiety with age statistically controlled?" The full model to answer this question is the same as the restricted model for the interaction question.

(Model 7-1) $Y_1 = a_0U + a_1X_1 + a_2T_1 + a_3T_2 + a_4T_3 + a_5A_1 + a_6A_2 + a_7A_3 +$

and the restricted model is:

(Model 7-2) $Y_1 = a_0U + a_1X_1 + a_2T_1 + a_3T_2 + a_4T_3 + E_3$

If the F test between Model 7-1 and Model 7-2 is significant, then the knowledge of level of anxiety does improve the prediction of the criterion and the treatments do have different effects at different levels of anxiety.

If the F test is not significant, the next question to be asked may be stated: "Do the treatments have an influence on the criterion variable with age statistically controlled?" The full model used in answering this question is:

$$Y_1 = a_0U + a_1X_1 + a_2T_1 + a_3T_2 + a_4T_3 + E_3$$

The restricted model is:

$$Y_1 = a_0U + a_1X_1 + E_4$$

If the F test is significant, the treatments do have a differential effect on the criterion measures in addition to the effect of the age of the individual. If the results are not significant, then we can answer a basic question asked earlier in the following way: The additional knowledge of treatment group and anxiety level does not significantly improve the researcher's ability to predict the test score of an individual, given the age of the individual.

The above question has dealt with only one concommitant variable. In the multiple linear regression model, it is quite easy to deal with more than one concommitant variable. Let us assume that in the problem being discussed, the researcher has available a pre-treatment test score. If the researcher wants to statistically control pre-treatment test score along with the age of the individual, the researcher simply adds a vector containing the pre-treatment scores to all of his models. Furthermore, the researcher may also want to statistically control the interaction that exists between the two control variables -- age and pre-treatment score -- which can be done by adding a product vector (age * pre-treatment score). As is well evident, the authors could continue this discussion indefinitely. The following model might be beneficial in reviewing the technique presented:

$$Y_1 = a_0U + A + B + E_1$$

where:

a_0 = regression constant.

A = represents all control variables and their relationships to each other.

B = represents all independent variables and their relationships to each other.

$E_1 = (Y_1 - \widetilde{Y}_1)$.

It can be noted that information in A is held constant in this problem, and the restrictions are made with respect to the variables represented by B. It is quite possible that the information to be included in the A segment of the equation could well be the result of previous computations. For instance, the researcher might want to decide if he should hold constant the interaction between age and pre-treatment score. He would simply test to determine if the interaction between the two control variables is significant such that it should be included in the study. This analysis should be done in the absence of the variables represented by B.

This technique provides an opportunity to the researcher to explore an infinite number of possibilities; but the sound researcher bases his models on some theoretical model and explores with respect to that model. He includes a concommitant variable in his research model only when he has reason to believe that it is acting as a predictor for the criterion being studied.

Relationship of Multiple Linear Regression to Various Multivariate Statistical Techniques

In previous chapters we have indicated the relationship of the multiple linear regression technique to various statistical techniques.

We have shown how one may answer the same questions as one does with the analysis of variance approach (from one-way analysis of variance to analysis of variance designs employing more than one independent variable). We have also discussed the applicability of the multiple regression technique to curvilinear questions, as well as to questions involving the control or consideration of confounding variables.

In all of the regression models presented in this text, we have considered only one dependent variable. The majority of the regression models did contain multiple independent variables, but a single criterion variable was always employed. Are we, however, always sure that a particular variable is a good measure of the criterion behavior that we are trying to investigate? When studying college success, is grade point average the only aspect of college success that we want to consider important? Should other variables be taken into consideration, such as number of job offers and personal satisfaction? Or, to follow a line of reasoning consonant with the multiple linear regression approach: Is there an optimum weighted linear combination of dependent variables, as well as an optimum weighted linear combination of independent variables?

The recent trend of many research areas has been to admit that there is often no single good measure of the criterion measure under consideration. The criterion may even be multi-dimensional and thus impossible to measure adequately with a single variable. The statistical procedures which consider multiple criterion variables are called multivariate techniques, and this definition shall be used throughout.

We shall soon discover that the multiple linear regression technique is a restricted case of some of these multivariate techniques, just as we have already discovered that the bivariate correlation is a restricted case of the multiple linear regression model.

Multiple Correlation

In actual practice there is no difference between multiple correlation and regression analysis. There is, though, a slight theoretical difference. With multiple correlation, we allow the subjects to obtain any value on the predictor variables; we, in effect, have a "random effects" model. In regression analysis, the researcher actually chooses the values that the subjects will obtain on the predictor variables. For instance, only two drug levels, three methods of teaching, or two sexes are studied. In this latter case, generalizations of the results can be made only to those "fixed" values originally chosen by the experimenter. In the multiple correlational model, generalizations can be made across the entire range of values studied. One should note that in many cases when we have applied the multiple linear regression approach, especially to continuous predictor variables and in the previously discussed covariance model, that we are actually applying the multiple correlational model.

We can turn to several developments with the multiple correlational model to gain some further insights into the multiple regression model. These developments involve the interpretations of beta weights, shrinkage formula, judicious choice of predictor variables, and the associated concept of suppression variables.

The interpretation of beta weights makes sense when we are dealing with continuous variables. The R^2 is the proportion of criterion variance that is accounted for by the set of predictor variables. The systematic determination of the most important set of variables can be accomplished by setting the partial regression weight of each variable equal to zero (which effectively eliminates that variable from the variable pool). The researcher may prefer to determine the amount of variance accounted for by each individual predictor variable without going through this process. The total amount of criterion variance associated with each variable is the product of the beta weight (B_i) times the correlation between the variable and the criterion (r_{ic}) or:

(Total variance associated with variable "i") = $B_i * r_{ic}$

The beta weights are given in the computer output for each model and are labeled "standard weights."

If the predictor variables are intercorrelated, which is usually the case, some of this total variance is shared with other predictor variance. To find the variance in the criterion which is solely associated with a particular variable (i), all we have to do is to square the beta weight, or:

(Criterion variance solely associated with variable "i") = $B_i * B_i$

Thus the total variance associated with the variable "i" is the sum of the criterion variance solely associated with variable "i" and shared variance, or:

(Shared variance associated with variable "i") = $(B_i * r_{ic}) - (B_i * B_i)$

The above line of reasoning only holds true when the products ($B_i r_{ic}$) are all positive. There are many other alternative ways of interpreting beta weights and these are rather lucidly discussed by Darlington in a 1968 issue of Psychological Bulletin. We would like to emphasize that any interpretation must be made in light of the other predictor variables and in light of the nature of the subjects.

The "correction for bias" or "shrinkage formula" discussed in most presentations of multiple correlation (Guilford, 1965) is a statistical approach to the problem of cross-validation. The present authors prefer a cross-validation procedure which employs actual data, for this will give an empirical check on the weighting coefficients, rather than a statistical estimate. One can seldom completely meet the assumptions of the statistical procedure, and the statistical estimate may be a biased estimate under those conditions.

Judicious choice of predictor variables is a concern to researchers for various reasons. The most important reason involves finances. We can increase the amount of variance accounted for by adding variables to the predictor set. But is the increase in the R^2 value worth the money, time, and effort involved in obtaining the additional predictor variables? And of even more importance, does the use of these particular predictor variables make theoretical sense? Whenever additional predictor variables are added, the data is overfitted until the number of predictor variables is equal to the number of subjects, wherein the R^2, of necessity, is one. We are accounting for all of the criterion variance with no restriction on the nature of the predictor variables employed -- the only stipulation being that we have as many predictor variables as subjects. Clearly this is not a very desirable situation; we should have some basis upon which to select predictor variables.

The present authors contend that selection of predictor variables should rest on an individual's theoretical orientation. But it must also be acknowledged that often there are several variables which measure the same construct.

> Given this state of affairs, the general rule-of-thumb is that the better predictor variable (in the sense of increasing R^2) is one which:
>
> 1. has low correlations with other predictor variables.
>
> 2. has high correlation with the criterion.

As with most rules-of-thumb, there is one basic exception, this exception being referred to as a "suppressor variable" or "suppression variable."

> The term "suppressor variable" is applied to a predictor variable which increases the R^2 but has the following properties:
>
> 1. extremely low (near zero) correlation with the criterion.
>
> 2. high correlation with another predictor which is highly correlated with the criterion.

A suppressor variable is completely antithetical to the concept of a "good predictor." After reading the discussion of factor analysis in this chapter, you may more fully understand how such a variable can increase the amount of variance accounted for in the criterion when, in fact, it is not correlated with the criterion. We will, therefore, present a numerical example at that point.

Multiple Discriminant Analysis

Consider a linear regression model with all predictor variables continuous and the criterion variable being a dichotomous vector indicating group membership. (See Equation 7.1.) With this model we are attempting to discover some weighting coefficients for each of the continuous vectors such that we can maximally predict the dichotomous criterion of group membership. Maximal prediction in this case involves generating predicted scores such that the predicted scores of the two groups are maximally different. We would test this difference by looking at the group means and the variability about the group means. What we are thus doing is maximizing the t-test for the difference between two means, or in general analysis of variance terms, we are maximizing the ratio of the between groups variance to the within groups variance. Let's take a closer look at the regression model which answers this question.

(7.1) $$Y_1 = a_0U + a_1X_1 + a_2X_2 + a_3X_3 + E_1$$

where: X_1, X_2, and X_3 are continuous predictor variables;

Y_1 is the dichotomous vector indicating group membership; and

a_1, a_2, and a_3 are the partial regression weights to be calculated.

Note: The standard partial regression weights are sometimes given a special name in the discriminant model -- "discriminant weights."

The regression weights are computed in the usual manner, and the amount of variance accounted for in this model (R^2) can be tested against an R^2 of zero. If there is a significant difference between the two R^2 values, then the two groups can be differentiated or discriminated with respect to these particular variables.

We could accomplish the same statistical outcome by applying a one-way analysis of variance design on the predicted scores (called "discriminant scores" in this application). Again, we note the overlapping of the two statistical models, for the greater the mean differences on the predicted scores (assuming that the within groups variability remains constant), the higher the differential predictability.

Conversely, if we have no differential predictability in Equation 7.1 ($R^2 = 0$), then there is no difference in group means, and the analysis of variance procedure will give us an insignificant F ratio.

The multiple linear regression model, though, can handle only the two-group multiple discriminant problem. If we attempt to indicate group membership for three groups with only one vector, we could use a "1" for Group A, a "2" for Group B, and a "3" for Group C. But this demands an ordered relationship between the groups -- a relationship that we probably do not want to assume. This is the same problem that we discussed at the end of Chapter VI with repsect to the senselessness of placing categorical groups in an ordered relationship.

If we have more than two groups for a criterion measure, the multiple linear regression model is no longer applicable, and we must turn to the multiple discriminant procedure. The research question involves determining the extent and manner in which two or more previously defined groups of subjects may be differentiated (or discriminated between) by a set of variables operating together. With two groups of subjects this separation can be represented only along a single dimension, but with more than two groups the differentiation may be described in terms of multiple independent dimensions. The maximum number of these "factors" or reference axes necessary to represent group differences will be the smaller of two numbers: the number of groups minus one, or the number of variables.

Tests of significance are available for both the overall group discrimination, and the discriminating ability of each of these reference axes (Veldman, 1967). Further interpretive information can be gained by computing discriminant scores for each person on each discriminant function. A discriminant score is the sum of the products of the discriminant weights times the individual's z scores. Correlating these new discriminant scores with each of the original variables will indicate which variables are more closely aligned with the reference axis, or what construct the discriminant axis is measuring. An application of the discriminant procedure may help clarify the process.

McNeil (1968) hypothesized that certain subcultures responded differently to a semantic differential instrument. The semantic differential, developed by Osgood and his associates (Osgood, Suci, and Tannebaum, 1957), purports to measure connotative meaning and thus may be considered to be measuring one's frame of reference.

Some researchers hypothesized that all people have the same basic frame of reference. The latter hypothesis did not seem tenable and was thus put to a statistical test.

Since the concept of "frame of reference" is a multi-dimensional construct, six measures of one's frame of reference were computed from the semantic differential data, and these scores were treated as the predictor variables. The total sample of subjects consisted of four subcultural groups: Middle class Anglos; lower class Anglos; lower class Negroes; and lower class Latin Americans. The results of the discriminant analysis appear in Figure 7-1.

	Discriminant Functions		
	1	2	3
Middle Class Anglos	.51	-.02	-.11
Lower Class Anglos	.20	-.07	.15
Lower Class Negroes	-.27	.56	.00
Lower Class Latin Americans	-.48	-.21	-.04

NOTE: Wilks' lambda = .780
$F = 7.407$ df = 18/df = 1449 $p < .0001$

First discriminant function 70.03% variance

$\chi^2 = 88.434$ df = 8 $p < .0001$

Second discriminant function 26.05% variance

$\chi^2 = 34.691$ df = 6 $p < .0001$

Third discriminant function 3.92% variance

$\chi^2 = 5.376$ df = 4 $p > .26$

Figure 7-1. Group centroids on the three discriminant functions.

Three discriminant functions were computed because the number of groups minus one is equal to three (while there are six predictor variables). Looking at the bottom of Figure 7-1, we notice that Wilks' lambda is equal to .780 and is highly significant. The three discriminant functions are together accounting for a significant (non-chance) amount of variance. Of the total amount of variance accounted for by the three discriminant functions, we note that the first discriminant function is accounting for over 70 per cent of the predicted variance. The approximate chi-square test indicates that this function is highly significant, or that the discriminant scores for these four groups on this function are more different than would be expected by random sampling. We also note that there is a second, independent (completely different) way of separating these four groups on the six dependent variables. The group means on each of these discriminant functions are indicated in Figure 7-1. Since the third and last discriminant function is not statistically significant, we can ignore that function with little loss of generality. The group means (group centroids) may be more meaningfully understood by plotting the results on a coordinate axis as in Figure 7-2.

Figure 7-2 reinforces the statistical conclusions that the four subcultures are different on not just one dimension, but on a second dimension as well. Further interpretive aids can be gained from inspection of the correlations between the computed discriminant scores and the original variables. Figure 7-3 contains this information, and we note that predictor variables three and four are not

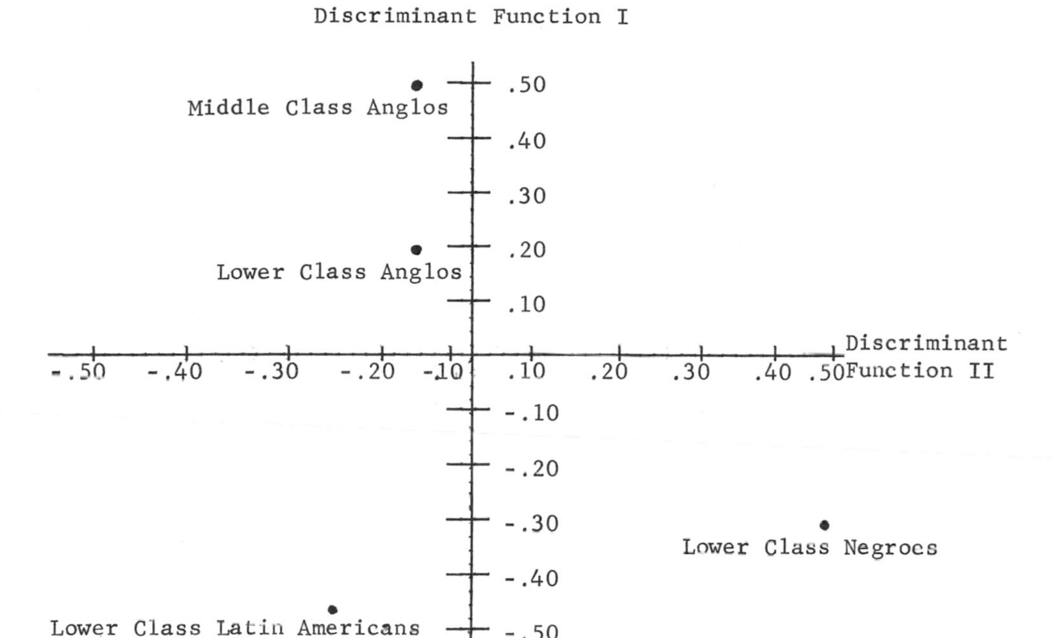

Figure 7-2. Graphic representation of discriminant results (from Figure 7-1).

	Discriminant Scores on Discriminant Function I	Discriminant Scores on Discriminant Function II
Predictor Variable 1	.61	-.37
Predictor Variable 2	.43	.60
Predictor Variable 3	.11	-.52
Predictor Variable 4	-.22	.45
Predictor Variable 5	-.46	-.07
Predictor Variable 6	.43	.21

Figure 7-3. Correlations between generated discriminant scores and the original six predictor variables.

related to this first discriminant function, whereas the second discriminant function is measuring what predictor variables two and four have in common with the opposite of dependent variable three. That is, a subject having a high discriminant score on the second discriminant function would most likely have high scores on predictor variables two and four and a low score on predictor variable three.

Factor Analysis

The technique of factor analysis can be considered a multivariate technique in the sense that the interrelationships of a number of variables (independent or dependent) are investigated simultaneously. There are a number of factor analytic techniques, but we are grouping all of them together for the purposes of this discussion.

The technique essentially operates on a correlation matrix and determines which variables are measuring the same underlying constructs, and to what degree these variables are measuring these constructs. The most widely used factor analytic techniques demand that the underlying constructs be independent of each other and that they account for completely different kinds of variance in the domain of behavior being studied.

If the domain of behavior being investigated (defined by the totality of the measures obtained) is unidimensional, then only one underlying construct (or "factor") will emerge. But if there are two or more underlying unrelated aspects of the domain, then two or more factors will be generated by the statistical procedure. If only some of the total variance in the behavior can be explained by one construct, while other constructs are necessary to explain (or account for) other amounts of the total variance, then our domain of behavior must be considered to be multidimensional.

Results from a factor analysis can indicate the extent to which a domain of behavior is multidimensional. We can also discover to what extent variables are overlapping in the kinds of variance that they are accounting for. Also, a variable may be highly unrelated to other variables, but still account for a certain amount of the variance in the behavior, i.e., the variable is accounting for some variance that no other variable is accounting for.

By now you may have an idea of some of the advantages of the factor analytic technique as it relates to multiple correlation. We have already discussed the ideal properties of a new predictor variable when adding it to a previously defined predictor set. Again, low intercorrelations with other predictors are desired, as well as a high correlation with the criterion.

Guilford presents this idea very convincingly (Guilford, 1965). Let us assume that we have already obtained our factor analytic results and that the correlations of some selected variables with the underlying construct appear as in Table 7-1. The original intercorrelation matrix of these four variables appear in Table 7-2.

Variable	Common Factors (Underlying Constructs)				
	A	B	C	D	F
Test 1	.60	.00	.60	.00	.00
Test 2	.40	.00	.35	.00	.80
Test 3	.00	.70	.00	.50	.00
Test 4	.00	.00	.00	.00	.80
Criterion J	.40	.30	.40	.50	.00

Table 7-1. Correlations (factor loadings) of experimental variables with the common factors (underlying constructs).

Variables	Test 1	Test 2	Test 3	Test 4	Criterion J
Test 1		.45	.00	.00	.48
Test 2	.45		.00	.64	.30
Test 3	.00	.00		.00	.46
Test 4	.00	.64	.00		.00
Criterion J	.48	.30	.46	.00	

Table 7-2. Intercorrelation matrix of four tests with Criterion J.

From Table 7-2 we note that the (multiple) correlation between Test 1 and the criterion J is .48. We can now ask the question, "Which variable would be the best to add to the predictor set, Test 2 or Test 3?" By looking at Table 7-2, we note that Test 3 has low intercorrelations with Test 1, and a relatively high correlation with the criterion (r = .46). Test 2, on the other hand, correlates .45 with the predictor variable already in use and only .30 with the criterion. Thus, we would probably prefer to add Test 3 to the predictor set of variables. Table 7-1 leads us to the same comclusion as does the above line of reasoning. We note that the criterion variable is related to four of the five common factors. We should also recollect that these common factors as described here are accounting for different kinds of variance. Therefore, 16 per cent ($.40^2$) of the variance in the criterion score is accounted for by Factor A, while 9 per cent ($.30^2$) is accounted for by Factor B, etc. We note that Factor F, while highly related to Test 2, is unrelated to the criterion. We also note that Test 2 is related to the same factors as is Test 1 (with the one exception of Factor F). Conversely, Test 3 is related to different factors than are related to Test 1, but these factors that Test 3 are related to are important sources of variance as far as the criterion score is concerned. That is to say, Test 3 accounts for sources of variance in the criterion which are not accounted for by Test 1 (or Test 2 as far as that goes).

The multiple correlation between the criterion and Test 1 and Test 2 is .46, whereas the multiple correlation between the criterion and Test 1 and Test 3 is .66. Thus we would account for more variance if we add Test 3 rather than Test 2 to the already defined predictor set of Test 1.

As we have indicated earlier, the concept of suppressor variable can be best understood through the concept of factor analysis. Let us suppose that we have a predictor set of Test 1 and 2 for the criterion J. We note from Table 7-2 that Test 4 is uncorrelated with the criterion (r_{4j} = .00), but that it has a high correlation (r_{42} = .64) with a variable (Test 2) which is itself correlated (r_{2j} = .30) with the criterion. These two properties (see page 233) lead us to believe that Test 4 may function as a suppressor variable and thus increase the magnitude of R when added to the already established predictor set.

The already established predictor set can be written as the following regression model which yields an R of .46 and associated R^2 of .21:

$$\mathcal{J}_1 = a_0U + a_1 \text{ (Test 1)} + a_2 \text{ (Test 2)}$$

The new predictor set can be written as the following regression model:

$$\mathcal{J}_1 = a_0U + a_1 \text{ (Test 1)} + a_2 \text{ (Test 2)} + a_3 \text{ (Test 4)}$$

The above model yields an R of .499 and an R^2 of just under .25. Test 4, which is uncorrelated with the criterion can, in fact, increase the amount of variance that the predictor set can account for in the criterion. If we look closely at Table 7-1, we can possibly understand how this happens. Test 4 and Test 2 have only one common underlying relationship -- that on Factor F. The criterion variable does not measure this underlying construct, so Test 4 effectively subtracts out from Test 2 this variance attributable to Factor F. This subtraction process may be more fully understood by investigating the beta weights associated with the predictor variables:

Variable	Beta Weight
Test 1	.38
Test 2	.22
Test 4	-.14

The suppressor variable (Test 4) thus has a negative beta weight associated with it. This example should serve to reinforce the reader's awareness of the complexity involved when multiple variables are included in a single analysis. If a researcher used a single variable approach, though, the influence of Test 4 would most likely never be discerned. Thus, you see once again the advantage of a multivariable research approach.

Canonical Correlation

The final statistical technique that we will discuss in this chapter is that of canonical correlation. This is fitting in that everything that we have previously discussed can be couched within the canonical model. The canonical model allows for a number of continuous and/or dichotomous criterion variables as well as a number of continuous and/or dichotomous predictor variables. Figure 7-4 indicates the relationship of the various techniques. Because of the generality of the canonical model, the regression model, and the Pearson Product model, these three models are applicable to any condition which falls below them in Figure 7-4. Thus, the canonical model can be applied to any situation in Figure 7-4. The regression model is a restricted instance of the canonical model but is itself applicable to situations "I" through "P" (and to "E" through "H" with appropriate interpretation). Included in Figure 7-4 are the comparable analyses of variance terms which yield the same conclusions as do the correlational procedures.

The canonical correlation model finds two sets of weighting coefficients, rather than just one set as in the linear model. Since there is not just one criterion variable, the procedure must find weighting coefficients so as to maximally predict a composite criterion score. The function that is minimized in multiple linear regression is the error sum of squares between the predicted criterion and the actual criterion. The function that is minimized in the canonical model is the error sum of squares between a predicted score (predicted from the one set of variables) and another predicted score (predicted from the other set of variables).

If one were to submit the two sets of predicted scores to the Pearson Product correlational model, one would obtain the numerical value of the canonical correlation (R_c). Thus, the canonical

Predictor Variable	Method: "Correlational"	Method: "Analysis of Variance"	Criterion Variable
A. Multiple Continuous	Canonical		Multiple Continuous
B. Multiple Dichotomous	Canonical	Multivariate Analysis of Variance	Multiple Continuous
C. Multiple Continuous	Discriminant		Multiple Dichotomous
D. Multiple Dichotomous	Discriminant	Multivariate Analysis of Variance	Multiple Dichotomous
E. Single Continuous	Canonical		Multiple Continuous
F. Single Dichotomous	two-group discriminant	Multivariate Analysis of Variance	Multiple Continuous
G. Single Continuous	Canonical		Multiple Dichotomous
H. Single Dichotomous	two-group discriminant	Multivariate Analysis of Variance	Multiple Dichotomous
I. Multiple Continuous	Regression		Single Continuous
J. Multiple Dichotomous	Multiserial R	Factorial Analysis of Variance	Single Continuous
K. Multiple Continuous	Discriminant		Single Dichotomous
L. Multiple Dichotomous	Regression	Factorial Analysis of Variance	Single Dichotomous
M. Single Continuous	Pearson		Single Continuous
N. Single Dichotomous	Point biserial	t-test	Single Continuous
O. Single Continuous	Point biserial		Single Dichotomous
P. Single Dichotomous	Phi-coefficient	Chi-square	Single Dichotomous

NOTE: Regression model can be applied successfully on models E through P, with appropriate interpretation.

Figure 7-4. Interrelationship of various statistical techniques.

correlation is nothing more than the bivariate correlation between two predicted scores. The square of the R_c can be interpreted as the amount of variance in the one composite score that is accounted for by the other composite score.

To gain an insight into what these composite scores are measuring, we can look at the bivariate correlations of the original variables with the new composite scores. This process is somewhat similar to the process of interpreting factor loadings. Variables which are highly correlated with the new composite scores are measuring somewhat the same construct, and variables lowly correlated with the composite score are not measuring the same construct.

As in factor analysis and discriminant analysis, the one canonical correlation will probably not account for all of the variance in the data. Additional canonical correlations are computed with the restriction that the subsequent composite scores be uncorrelated with the previous composite scores. It is entirely possible that the two sets of variables are significantly related in two (or more) ways. If this is the state of affairs, the researcher should be aware of the situation and incorporate the fact into his theory.

Tests of significance exist for each canonical correlation and for the total amount of variance accounted for in the two sets of variables. But other tests of significance are lacking, partly because of the recent arrival of the technique, partly because of the complexity of the technique, and partly because of the unavailability of appropriate data and researchers who are interested in questions which can be answered by the canonical model. The authors hope that advances will be made with this technique so that it becomes as flexible as the regression model. Specifically, tests of significance for comparing one canonical model against another need to be developed. Horst (1961) has extended the canonical model to more than two sets of data, but the topic is too complex to discuss here.

The careful reader will note that in the discussion of the canonical model we have disregarded the terminology of predictor and criterion variables and have discussed them as "two sets of data." As far as the mathematical calculations are concerned, these distinctions never really have to be made in any model. Interpretations of the data do have to take these distinctions into account, though. This point can be further understood by looking at the bivariate correlational model involving variables X and Y. Since the correlation between X and Y is the same as the correlation between Y and X, we can predict Y from X as accurately as we can predict X from Y. In a specific case either one of these variables may be considered as the criterion (or dependent) variable within the confines of the research question.

Chapter VIII Special Considerations Regarding Multiple Linear Regression Analysis

This chapter brings together a number of issues which do not conveniently fit into the normal flow of this text. Some of the points considered relate to rather practical further uses of DATRAN, whereas other points of discussion relate to mathematical derivation of formulae. Each sub-topic should be viewed as a self-contained unit. The first four topics should be read, since they have direct bearing upon the use of program LINEAR. The remaining topics should be viewed as optional reading.

Missing Data

The statistical models presented in this text are based upon complete data on all subjects. Often in behavioral research, complete data cannot be obtained for all subjects on all predictor variables. If the sample size is large and subjects with missing data are few, we recommend excluding the subjects with missing data. This straightforward solution for dealing with missing data is economical but also might reduce the representativeness of the sample in relation to the population. For example, if one were collecting data over several days in a school and a number of students do not attend all sessions, the missing data might be due to a systematic bias, i.e., these students might be truants and potential dropouts. If these students were excluded from the study, then the population which the sample represents is <u>not</u> all students in the particular grade level but rather all the students in the particular grade level who attend during testing sessions.

In the case where one suspects a systematic bias among subjects with missing data, one may wish to consider the use of the procedure suggested below.

For each bit of missing data, determine the continuous observed scores that occur in the vector where the one bit is missing. From among these scores, randomly assign one score to the subject who is missing the score.

The random procedure might be conducted by putting each observed score for the variable under consideration on a separate card and then blindly selecting a card to determine the score value to be assigned for the missing element. If more than one subject is missing data for the same variable, the card can be replaced and randomized, and another card can be pulled to provide the score for the next subject with missing data. [For each bit of fictitious data inserted into the data set, you might want to reduce the value of N (where N is used to calculate degrees of freedom) by "1" in order to have a more conservative test.]

The procedure just recommended is preferred over the procedure of assigning the mean value of the variable to the missing points because the addition of a number of means artificially reduces the variance of the variable.

A test to determine whether the subjects with missing data are different from subjects with no missing data with respect to the criterion can easily be made. Create a new vector and place a "1" in this vector if the criterion score is from an individual with missing data and a "0" otherwise. Then in your full model include all predictor vectors plus your dichotomous missing data vector. Make your first restriction such that the weight associated with the missing data vector equals zero. If a significant decrease in variance accounted for is observed, then these subjects are likely to be different from your subjects without missing data.

Other procedures for generating missing data can be found in Snedecor (1946) and Cochran (1953).

Program LINEAR as an Iterative Solution

Program LINEAR does not yield an exact solution to the statistical problem, although the approximations arrived at through the iterative process are close enough for most research purposes. If more accuracy is desired, the stop criterion on the model card can be increased from E-05 to E-06 or even higher if more accuracy is deemed necessary. Increasing the stop criterion will generally allow for more iterations and thus more computer time. The increase in computer time is, for most practical purposes, negligible.

The authors have experienced difficulty with the iteration procedure. Even though the problem is of little practical importance, we feel that by discussing this problem, we will clarify the notion of the iteration process.

Whenever we make a restriction on a full model, we are losing some predictive information and thus increasing our error of prediction. The amount of criterion variance that we are accounting for in the restricted model is usually less than the amount of variance accounted for in the full model. The restricted R^2 will be less than the full model R^2; or if no predictive information has been lost, the restricted R^2 will be equal to the full model R^2. But it is not mathematically possible to have a restricted R^2 which is larger than the full model R^2.

Contrary to the previous discussion, in some LINEAR solutions we have found the restricted R^2 to be larger than the full model R^2. We believe the cause of this phenomenon can best be explained by Figure 8-1. In Figure 8-1 we have plotted the R^2 after each iteration for both the full and restricted models. It should be noted that there is a brief leveling of the full model curve at about the tenth iteration. The increase in the R^2 value from the tenth to the eleventh iteration may be less than the value of the stop criterion. If this is so, then the iteration procedure would stop at this point yielding an R^2 value of about .55.

A leveling for the restricted model does not occur until the twenty-fifth iteration, and if the stop criterion is satisfied, an R^2 value of about .70 (higher than the R^2 associated with the full model) would be outputted for the restricted model.

If the stop criterion were set higher (e.g., E-08 or E-12), the iteration process might be continued through forty or fifty iterations for both the full and restricted models. If this were the case, the full model R^2 would indeed be greater than the restricted model R^2.

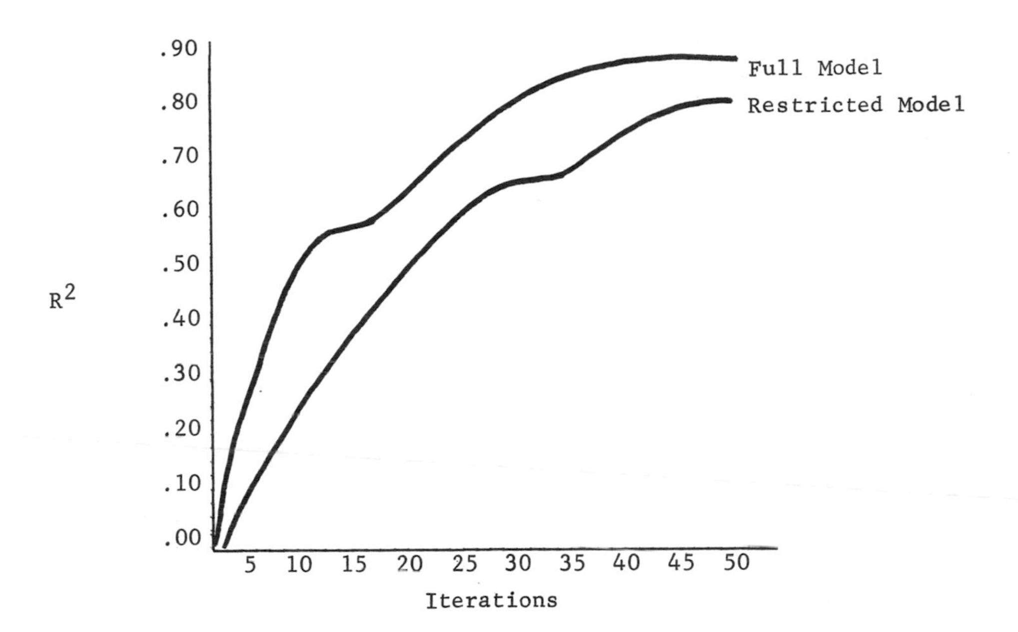

Figure 8-1. Amount of variance accounted for as a function of the number of iterations.

In practice, we have circumvented this problem in one of two ways: by either (1) increasing the stop criterion as mentioned above, or (2) by including in the full model all vectors which appear in the restricted model. By this second statement, we mean that it is good practice to include in the full model all linearly dependent vectors which appear in the restricted model. Suppose that we have two IQ vectors, one for males and one for females:

$$\tilde{Y}_1 = a_0U + a_1X_1 + a_2X_2 \quad \text{(Full model for iteration problem)}$$

Now, suppose that we want to make the restriction that $a_1 = a_2 = b_1$:

$$\tilde{Y}_1 = a_0U + b_1X_3 \quad \text{(Restricted model for iteration problem)}$$

where: $X_3 = X_1 + X_2$, or a vector of IQ scores.

In order to circumvent the problem under discussion, we would include in the full model the linearly dependent vector X_3:

$$\tilde{Y}_1 = a_0U + a_1X_1 + a_2X_2 + b_1X_3$$ (Alternate full model for iteration problem)

The above model is the same as the full model because X_3 is redundant information, but the alternate model will circumvent the iteration problem.

Let us close this discussion by indicating that this is not a problem of much practical importance because whenever this problem has arisen, we have found, with the suggested changes, that the amount of variance accounted for by the restricted model is very close to that of the full model. That is, the restricted model is not significantly different from the full.

Some Further Uses of DATRAN

A. Generating Group Membership Vectors from Discrete Data. Whenever the researcher has many group designations, it would be foolish for him to include a group membership vector for each group on each card. A more realistic procedure would be to numerically code the various groups on a single column or on two adjacent columns. Suppose that a research question involves six groups. The group membership codes (1, 2 . . . 6) are punched in the first columns of the data cards.

If we read with the format card the group membership vector and two other vectors (which might represent a pre-test variable and the criterion), the original number of variables read would be "3." We can now use DATRAN to generate the six group membership vectors (mutually exclusive categories) that we need. We do this by preparing the following cards to be placed in the LINEAR deck:

```
K=A(I+1)
                  All statements in DATRAN should start
L=J+K             in column 7.

A(L+3)=1.0
```

The first card identifies the vector we want to manipulate, our original group membership vector -- in this case, vector A(I+1). K can equal 1, 2, 3, 4, 5, or 6, depending on the punch found in that vector for a particular subject. The second card identifies a new entity, that of J+K, to be further manipulated by the computer. This means that in the last card, A(L+3) essentially equals A(J+K+3), if that looks any more familiar to the reader. For the generalized case, the "L+____" value will be the total number of vectors (original plus previously generated) used prior to this transformation. In this case it is 3.

For each subject the vectors A(J+1), A(J+2), and A(J+3) will be the three original vectors read into the computer from the data cards. The computer reads the data for one subject at a time. With these three transformation cards, six new vectors are read for each subject. If the punch in the first vector for a subject is "1," then:

since K=A(I+1) and A(I+1)=1, then K=1;

since L=J+K and K=1, then L=J+1;

since A(L+3)=1.0 and L=J+1, then A(J+4)=1.0.

Thus, a new vector (A(J+4)) has been created containing a "1" for all subjects having a "1" in vector A(I+1).

If the punch in the first vector for another subject is "4," then: K=4, L=J+4, and A(J+7)=1.0.

A new vector (A(J+7)) has been formed containing a "1" for all subjects having a "4" in vector A(I+1). You can follow this procedure to determine the other four generated vectors.

If the punch in the group membership vector for another subject is "4," then K=4 and A(J+7) for that subject equals 1. Since in this case vectors A(J+4) through A(J+9) were set to zero by subroutine ZEROST, a subject will have a "1" punch in only one of the six group membership vectors, the vector appropriate to his code in vector A(I+1).

There are now nine vectors (original plus generated) for this problem.

Another example may make this process clearer. Suppose the first three vectors read on a card are scores of some sort and the fourth vector contains group membership codes for seven groups. Then a DATRAN statement was used to make A(J+5)=A(I+1)*A(I+2). There are now five previously used vectors. To obtain the seven group membership vectors, we use the following DATRAN statements:

```
K=A(I+4)        4 = number of group membership vector
L=J+K
A(L+5)=1.0      5 = number of previously used vectors
```

The parameter card for this problem (N = 25), if no further data transformations are performed, will be:

000250000400012

<u>B. Using DATRAN to Generate Interaction Vectors</u>. Once you have generated group membership vectors as in the previous example, if you wish to create vectors which contain a score for members of a particular group, zero otherwise, you can do this by adding one more DATRAN statement:

A(L+M)=A(I+N)

where: M = number of previously used vectors;

N = number of the vector containing the scores you wich to have interacting with group membership.

In the first example of the previous section, there would be nine previously used vectors. We want to test for interaction between group membership and the scores in vector 2. The DATRAN statement would be:

A(L+9)=A(I+2)

This would create six more vectors. The parameter card would now read (for 75 subjects):

000750000400015

C. Generating Group Membership Vectors from Continuous Data. There are several instances wherein the present authors have found a need to derive group membership vectors from continuous data. The most obvious case is when we need to generate groups "above median" and "below median" or "above the mean" and "below the mean." Another similar application involves generating groups from multi-digit code numbers. For instance, all male subjects may have been coded above 600, whereas all female subjects have been coded as 500 or below. In this latter example, one DATRAN statement will generate the one sex vector:

IF(A(I+1).LT.600.0)A(J+5)=1.0

Thus, if the code number in A(I+1) is less than 600, then the vector A(J+5) will have a "1" placed in it. If the code is greater than 600, then the vector A(J+5) will not be disturbed and a "0" otherwise. If for interaction purposes, you want two sex vectors, a male vector can be generated with one more DATRAN statement:

```
IF(A(I+1).GT.599.0)A(J+6)=1.0
```

A(J+6) then will have a "1" punch if male and a "0" otherwise.

D. Printing and Punching Information Via DATRAN. Any valid FORTRAN statement can be inserted into DATRAN, and therefore, we may perform printing and punching operations within DATRAN just as we can within any other subroutine. We may, for instance, want to print the code numbers of those subjects missing a particular score. We may want to punch a predicted score or punch total scale scores from the subscales that were read in. We may want to generate some new data cards containing ratios, for the purposes of additional analysis.

Suppose that we desire to print the code numbers of persons having a criterion score above 60 and that we also want to punch some new scores, these new scores being ratios of several body measurements to the total height of the subject. These ratio scores will be needed for further analyses, and even though we could generate them each time we so desired, the new set of data cards containing these ratios will save us a lot of trouble and computer time.

Suppose that our data is in the following format:

columns	
1-3	code
4-5	criterion
6-7	body measurement #1

8-9	body measurement #2
10-11	body measurement #3
12-13	body measurement #4
14-15	body measurement #5
16-17	body measurement #6
18-19	body measurement #7
20-21	height (from toe to waist)
22-23	height (from waist to top of head)

We would read the data by the following format: (F3.0,10F2.0). Note that the code must be read in the floating point (F) format. <u>This demands that no alphabetic punches be in the code</u>! We cannot over-emphasize this point. <u>Always use numbers to code your subjects</u>.

The following DATRAN statements will accomplish our desired results:

```
 IF(A(I+2).GT.60)  PRINT6,A(I+1)
6 FØRMAT(F6.0)
 A(J+12)=A(I+10)+A(I+11)
 DØ 8 K=3,9
 L=J+K
8 A(L)=A(L)/A(J+12)
 PUNCH 9,(A(J+K),K=1,12)
9 FØRMAT(12F5.2)
```

<u>A Special Application of the Sine Function</u>

In special research situations where cyclical variation might approximate the sine curve (see Figure 8-2), a linear model can be constructed to fit the curve. Figure 8-2 shows a three-cycle curve (from A to B, B to C, and C to D). If each complete cycle is defined to be 360°, then two transformations can provide the necessary vector to represent the curve. First the degrees represented by the X-axis must be converted to radians (X specified in degrees times (6.283/360.0)), and then obtain the sine of the radians. These transformations, of course, can be performed by subroutine DATRAN. For example, if the first vector read into the computer was the degrees and the second was the criterion, the transformation cards would be:

```
A(J+3)=A(I+1)*(6.283/360.0)
A(J+4)=SIN(A(J+3))
```

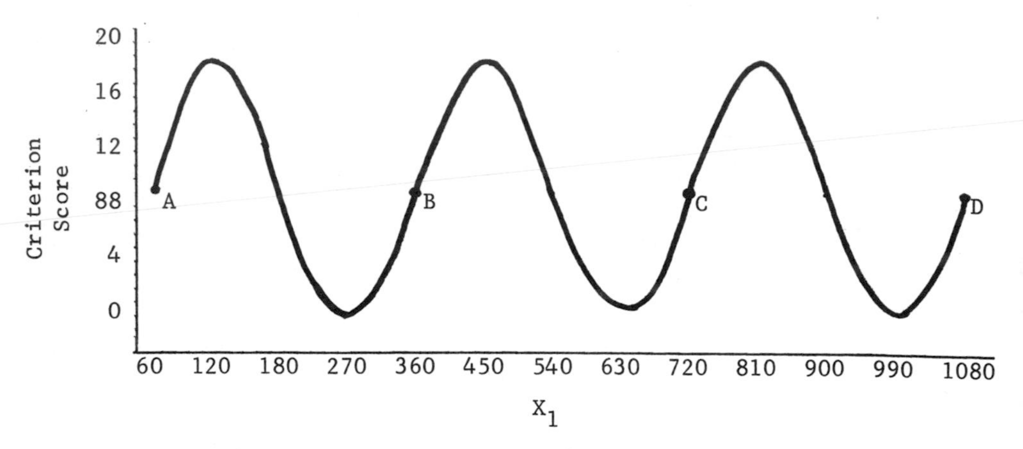

Figure 8-2. Three cycles represented, each composed of 360°.

Problem 8-1

In order to demonstrate a sine problem similar to the curve in Figure 8-2, the following data can be run:

```
SINE FUNCTION  0 TO 20
000420000200004
(F3.0,F2.0)
00011
01813
03614
03616
07220
09020
10820
14418
```

```
14416
16218
16213
18010
19807
21601
23401
27000
27000
32405
34208
36011
37813
41419
43220
45020
48618
50416
52214
54010
57600
61001
63000
64801
66603
70202
70208
72012
75617
77418
81020
81020
84619
88215
MODEL 1    005001.E-050101040402
MODEL 2    003001.E-05040402
F RATIO    006         010200010039
F RATIO    006         029000010040
```

The transformation cards are the same as those indicated above. Model 1 includes the degrees (X_1) and the sine of the radians (X_4) as the two predictor vectors. If the weight of X_1 is non-zero, a linear relationship would be indicated between X_1 and the criterion, and the X-axis would be rotated as in Figure 8-3. The problem given yields a zero weight for X_1, and if you plot the predicted scores, the figure takes the form as shown in Figure 8-2. (See the Answer Section for the computer output for this problem.)

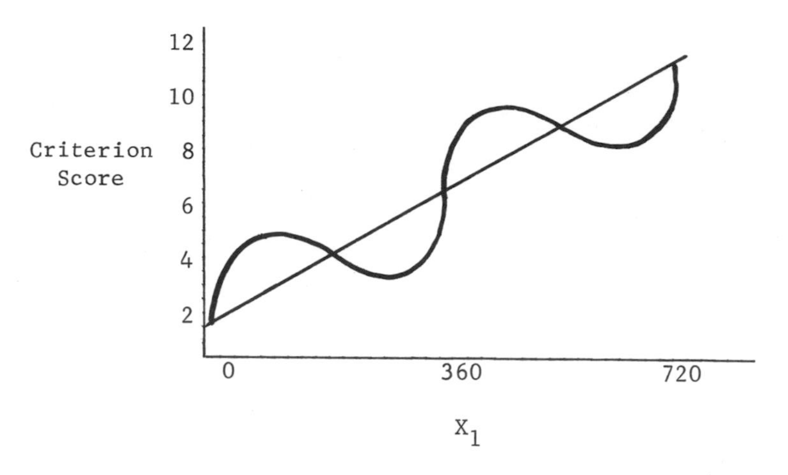

Figure 8-3. Two cycles represented with a non-zero weight associated with X_1.

Practical use of the sine function might include seasonal change in weather, economic behavior, etc., where each complete cycle (i.e., a year) can be divided into 360°. Other possible applications might include representing fixed interval reinforcement schedules, where the cycle might represent the interval between two reinforcing conditions.

If the sine curve represents only one cycle (e.g., from A to B in Figure 8-2), the sine curve can be approximated using three vectors as discussed in Chapter VI. This model would include: (1) the X_1 vector; (2) a vector whose elements are the square of the X_1 elements; and (3) a vector whose elements are the cube of the X_1 elements. Such a model will allow for the two points of inflection which characterizes a sine curve.

Computing an Eta Coefficient Via LINEAR

The eta coefficient indicates the maximum curvilinear relationship between two variables. In essence, each different value of the predictor variable is treated as a different "group." The variability of the criterion scores about each "group" mean is calculated as the measure of error. Consider the data in Figure 8-4.

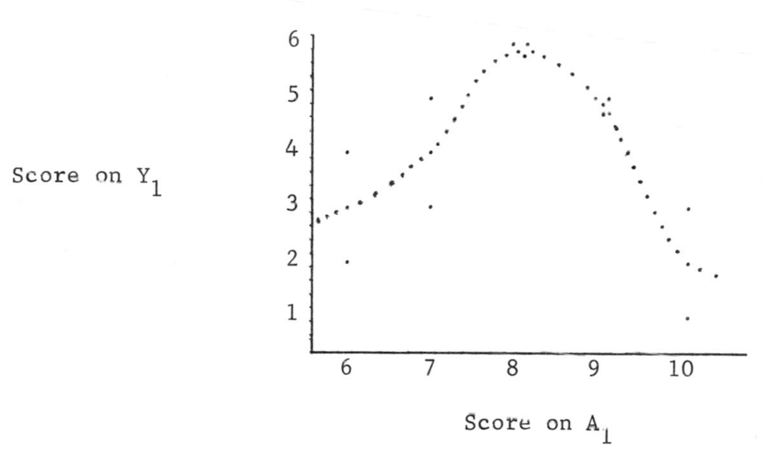

Figure 8-4. Schematic diagram of eta coefficient.

The "line of best fit" is depicted as falling on the mean of the criterion scores for each possible value of A_1. That is, the mean of Y_1 for all persons having a score of 6 on A_1 is 3. This value of 3 is used as the best estimate of the criterion score for all persons having a score of 6 on A_1. In order to depict the data in Figure 8-4 in terms of linear regression, we would construct the following full model:

$$Y_1 = a_0U + a_1X_1 + a_2X_2 + a_3X_3 + a_4X_4 + a_5X_5 + E_1$$

where:

Y_1 = the criterion.

U = the unit vector.

X_1 = 1 if person has a score of 6 on A_1; zero otherwise.

X_2 = 1 if person has a score of 7 on A_1; zero otherwise.

X_3 = 1 if person has a score of 8 on A_1; zero otherwise.

X_4 = 1 if person has a score of 9 on A_1; zero otherwise.

X_5 = 1 if person has a score of 10 on A_1; zero otherwise.

E_1 = the error vector.

$a_0, a_1, \ldots a_5$ are unknown weights.

The reader may wish to verify that the ESS associated with the above vector is 6. Once we have calculated the above full model, we can ask several questions. One important question that we can ask is: "Is this eta coefficient significantly different from zero?"

We can test this question by setting a_1, a_2, a_3, a_4, and a_5 equal to 0, essentially saying: "What is the error in prediction when I have no knowledge of a person's A_1 score?" The restricted model for this question will, of course, be:

$$Y_1 = a_0U + E_2$$

and will yield an R^2 value of 0. Note that the number of linearly independent vectors in the full model is equal to the number of different values of A_1.

Another important question is: "Is there a significant curvilinear relationship, over and above the linear relationship?" Note that if we make the following restrictions ($a_1 = a_2 = a_3 = a_4 = a_5 = a_6$, a common weight), we end up with the following model:

$$Y_1 = a_0U + a_6X_6 + E_3$$

where:

Y_1 = the criterion vector.

U = the unit vector.

$X_6 = X_1 + X_2 + X_3 + X_4 + X_5$, or simply the vector of A_1 scores.

E_3 = the error vector.

a_0 and a_6 are unknown weights.

But the above model has as predictors the unit vector and a continuous vector X_6; therefore, we will obtain from this model the linear relationship between A_1 and Y_1. Testing this restricted model against the full model will indicate whether there is a significant curvilinear relationship over and above the linear relationship.

We may generate the predictor variables necessary for the full model through DATRAN. Assume that the data is read as:

A(I+1) = criterion Y_1

A(I+2) = predictor A_1

The vectors X_1, X_2, X_3, X_4, and X_5 can be generated in the following manner:

```
K=J+(A(I+2)-3.0)
A(K)=1.0
```

A person who has an A_1 score of 6 (A(I+2)=6) will have a 1.0 placed in vector A(J+3), and all other vectors will be left with "0's" in them. Likewise, a person with a maximum A_1 score of 10 will have a "1" placed in vector A(J+7).

Solving Various Special Correlational and Analysis of Variance Problems with LINEAR

As discussed earlier, LINEAR is applicable to several special techniques. By indicating these applications, we have tried to illustrate the flexibility of the multiple linear regression approach.

The special F test formulae presented for testing the question of an R different from zero is a restricted case of the general F test that we have been using all along. Also, the special formula for testing the significance of beta weights can be derived from the general F test.

Jennings (1965) has discussed the techniques of part and partial correlation in terms of the linear regression model. Solving a "correlated t" or finding the difference between two (correlated) correlation coefficients is also feasible with the multiple regression approach.

But let us not lose sight of our goal. We want the statistical technique to fit our research question rather than vice versa. It is comforting to know that a "generalized technique" can cope with these various techniques that for some reason or another have been given special names, but it is more encouraging to know that one can build his own statistical model once he has specified his research question.

Let us end this section with a specific example of a feasible research question which would be difficult for the researcher to find an analysis of variance model to answer.

Suppose that we want to know if males are significantly more than three units higher than females on an aptitude test. That is, we are not going to be concerned about sex differences, unless the differences are greater than 3 units. The full model which contains information about sex is:

$$Y_1 = a_0U + (a_1+3)X_1 + a_2X_2 + E_1 \quad \text{(Full model for 3-unit question)}$$

where:

Y_1 = the achievement criterion

U = the unit vector

X_1 = 1 if person is male; zero otherwise

X_2 = 1 if person is female; zero otherwise

E_1 = the error vector

a_0, a_1, a_2 are unknown weights

Now we can simplify the full model by expanding the coefficient associated with X_1:

$$Y_1 = a_0U + a_1X + 3*X_1 + a_2X_2 + E_1$$

Since $3*X_1$ is a known quantity, we can transfer this quantity to the left side of the equation to get:

$$(Y_1 - 3*X_1) = a_0U + a_1X_1 + a_2X_2 + E_1$$

What we are effectively doing is subtracting a numerical value of three from the criterion score of each person in group X_1 (all males). We are not ready to make the usual restriction: $a_1 = a_2$.

$$(Y_1 - 3*X_1) = a_0U + E_2 \quad \text{(Restricted model for 3-unit question)}$$

If the restricted model is significantly less predictive than the full model, then we know that the restriction is not valid. We must be careful in interpreting the results, though, because our restriction said that the (adjusted) mean of the males was equal to the females. In order to infer that the males are more than three units better than the females, we must investigate the actual group means. That is, the significance could be due to the fact that the girls are much higher on the criterion than are the boys. We are actually concerned here with a one-tailed test and as such should adjust the probability levels associated with the tabled F values (which are conventionally two-tailed values).

Conventional Analysis of Variance Via Multiple Linear Regression Analysis

We have indicated previously in Chapter IV the relationship between analysis of variance and linear regression. In Chapter VI we have discussed the solution of the analysis of variance "interaction" question in linear regression terms. We would now like to discuss the conventional solution to the "main effects" question, both with equal N, proportional N, and unequal N.

Most statistical texts indicate that if a researcher finds significant interaction, then the "main effects" test is not appropriate. We agree with the intent of this research approach, but we argue that one should look at the nature of the interaction. It could very well be that the interaction is of the nature as in Figure 8-5.

If the researcher is forced to make a recommendation as to what treatment to administer to which children, he would unhesitatingly administer Treatment I to all children (assuming the two treatments are alike with regard to time, cost, etc.).

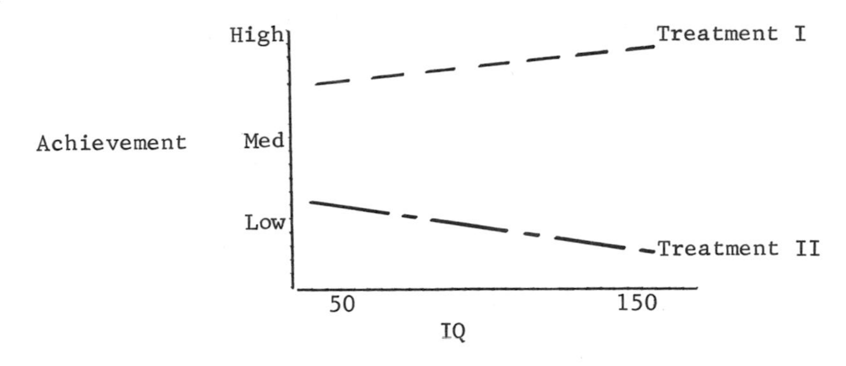

Figure 8-5. An interacting state of affairs, where the parallel lines do not cross within the range of interest.

If the interaction takes the form as in Figure 8-6, then other remarks are in order. That is, Treatment I should be given to high IQ students, while Treatment II should be given to low IQ students.

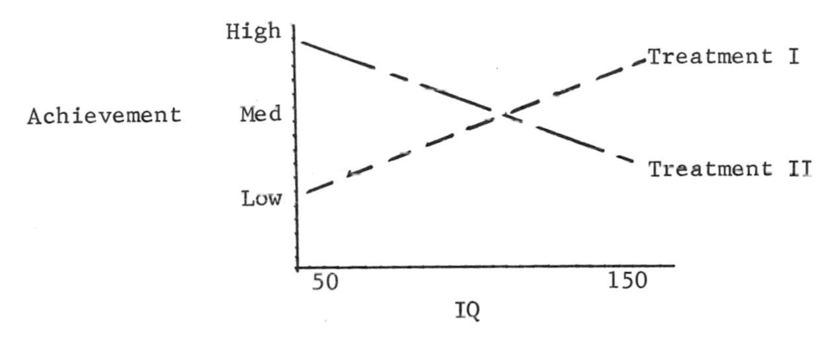

Figure 8-6. An interacting state of affairs, where the parallel lines cross within the range of interest.

When a researcher does not find significant interaction in the conventional analysis of variance procedure, then he may turn to the "main effects" question and treat the main effects as representative of the "simple effects" at each level. Statistical authors disagree on the error term that should be used for the main effect tests.

Some authors suggest using the "within" term, the same error term used for the interaction question. This term, though, assumes that there is no interaction effect, whereas the interaction test has only indicated that there is not enough interaction variance to worry about (i.e., the amount of interaction variance present is not significant).

Another approach to the problem of which error term to use is the process of "pooling." What is essentially done is that the sum of squares due to interaction is combined with the sum of squares due to the within group variability. This new sum of squares is divided by the sum of the two appropriate degrees of freedom. For an illustrative example, we will utilize an example presented by Spence, _et al_. (page 182). Table 8-1 gives the source table for the hypothetical experiment. The reader will note that the within group mean square (MS) was used as the error term for all three F tests. The $F_{2,24}$ for the interaction was computed as: 12.40/16.77 = .76. The $F_{1,24}$ for the affectivity main effect is: 158.70/16.27 = 9.76. The $F_{2,24}$ for the familiarity main effect is 81.74/16.27 = 5.02. The only significant effect ($p < .05$) was the affectivity main effect.

Source of Variation	SS	df	MS	F	P
Familiarity (A)	163.47	2	81.74	5.02	$<.05$
Affectivity (B)	158.70	1	158.70	9.76	$<.01$
Interaction (A*B)	24.80	2	12.40	.76	$>.05$
Within Groups (w)	390.40	24	16.27		
TOTAL	737.37	29			

Table 8-1. Summary of the analysis of variance of experiment on perceptual thresholds as related to word familiarity and affectivity.

In this example the authors did not choose to pool their error term. We will first indicate how one can reproduce the results in Table 8-1 through the linear regression approach, and then we will indicate how the "pooling" process in the calculation of the main effects can be accomplished.

Reproduction of Interaction Effects Via Linear Regression

The model which allows each group (cell) to have its own mean is:

(Model 8-1) $Y_1 = a_0U + a_1X_1 + a_2X_2 + a_3X_3 + a_4X_4 + a_5X_5 + a_6X_6 + E_1$

where:

X_1 = 1 if subject were a member of high neutral; 0 otherwise.

X_2 = 1 if subject were a member of medium neutral; 0 otherwise.

X_3 = 1 if subject were a member of low neutral; 0 otherwise.

X_4 = 1 if subject were a member of high unpleasant; 0 otherwise.

X_5 = 1 if subject were a member of medium unpleasant; 0 otherwise.

X_6 = 1 if subject were a member of low unpleasant; 0 otherwise.

The interaction question asks how reasonable the following restriction is:

$$a_1 - a_2 = a_3 - a_4 = a_5 - a_6$$

A number of equally good restricted models emanate from the above restriction. One that is most conceptually appealing to us is:

(Model 8-2) $Y_1 = a_0U + a_7X_7 + a_8X_8 + a_9X_9 + a_{10}X_{10} + a_{11}X_{11} + E_2$

where:

X_7 = 1 if member of high familiarity; 0 otherwise.

X_8 = 1 if member of medium familiarity; 0 otherwise.

X_9 = 1 if member of low familiarity; 0 otherwise.

X_{10} = 1 if member of emotionally neutral; 0 otherwise.

X_{11} = 1 if member of emotionally unpleasant; 0 otherwise.

Comparing Model 8-2 against Model 8-1 will yield the same F value as in the conventional analysis. Note that df_1 for the comparison is equal to 6-4, or 2, and df_2 is equal to N-6, or 24. These values correspond to those in Table 8-1.

Reproduction of Main Effects Via Linear Regression -- Without Pooling

We will proceed with the main effect for affectivity, without pooling the variances. What we mean by "main effect" is: $\mu_{EN} = \mu_{EU}$ (i.e., is the population mean of the "emotionally neutral" equal to the population mean of the "emotionally unpleasant"?). We can reconstruct estimates of these two means from our data via the following:

$$\frac{Na_1 + Na_2 + Na_3}{N + N + N} = \frac{Na_4 + Na_5 + Na_6}{N + N + N}$$

If we multiply both numerator and denominator by (N + N + N)/N, we get:

(8.1) $a_1 + a_2 + a_3 = a_4 + a_5 + a_6$

This is our restriction for the main effects question with respect to the affectivity effect. Note that we are making one restriction on our full model (Model 8-1). We will thus have five linearly independent vectors in our "main effects" restricted model. There are, of course, several equally valid restricted models, depending upon which value one solves for in Equation 8.1. Suppose that we solve for a_6: $a_6 = a_1 + a_2 + a_3 - a_4 - a_5$. Then the restricted model would be:

(8.2) $Y_1 = a_0U + a_1X_1 + a_2X_2 + a_3X_3 + a_4X_4 + a_5X_5 + (a_1+a_2+a_3-a_4-a_5)X_6$

(Model 8-3) $Y_1 = a_0U + a_1(X_1+X_6) + a_2(X_2+X_6) + a_3(X_3+X_6) + a_4(X_4-X_6) + a_5(X_5-X_6) + E_3$

Note that in Model 8-3 there are five vectors, plus the unit vector, yielding five linearly independent vectors. In testing Model 8-3 against Model 8-1, $df_1 = 6-5 = 1$ and $df_2 = N-6 = 24$. Again note that these degrees of freedom correspond to those in Table 8-1.

Note that in the above restriction we can reconstruct the two group means very easily even if the groups have an unequal number of subjects in each group. Let N_1 be the number of subjects in Group 1, N_2 the number of subjects in Group 2, etc. Then the restriction looks like:

$$\frac{N_1a_1 + N_2a_2 + N_3a_3}{N_1 + N_2 + N_3} = \frac{N_4a_4 + N_5a_5 + N_6a_6}{N_4 + N_5 + N_6}$$

This restriction does not simplify any further but can, of course, be substituted into the full model (Model 8-1). Again, we see the flexibility of the linear regression approach in that the requirement of proportionality is not mandatory.

The appropriate restrictions for the main effects question for familiarity would be:

(8.3) $a_1 + a_4 = a_2 + a_5 = a_3 + a_6$

Here we have two restrictions; therefore, when we test the restricted model against Model 8-1, we would have $df_1 = 6-4 = 2$ and $df_2 = 30-6 = 24$. The F yielded by this test would be numerically equal to that of Table 8-1 and the degrees of freedom would also be exactly the same.

Reproduction of Main Effects Via Linear Regression -- With Pooling

As has been mentioned previously, various conventional analysis of variance authors suggest pooling of error terms when a non-significant interaction is found. This process generates a "new error term" which is more stable than merely the within groups error term used alone.

If the data in Table 8-1 had been treated in the above fashion, the error term for both main effects would have been:

$$\frac{SS_w + SS_{A*B}}{df_w + df_{A*B}} = \frac{390.40 + 24.80}{24 + 2} = 15.97$$

Note that the "new error term" acquires a new value for the degrees of freedom (26).

The analogous process in linear regression is to make the following restriction on Model 8-2 when testing for "affectivity main effect": $a_{10} = a_{11}$.

That is, we are forcing a certain state of affairs on a model which has some error in it due to the slight (but non-significant) interaction between affectivity and familiarity. This state of affairs demands that there be no difference between the emotionally neutral category and the emotionally unpleasant category. The above restriction substituted into Model 8-2 yields the following model:

(Model 8-4) $Y_1 = a_0U + a_7X_7 + a_8X_8 + a_9X_9 + E_4$

The above model has three linearly independent vectors, and when tested against Model 8-2, has $df_1 = 4-3 = 1$ and $df_2 = N-4 = 30-4 = 26$. Note that these degrees of freedom again correspond to those yielded by the "new error term." The F value resulting from comparing these two models would be: 158.70/15.97.

For the sake of completion, the restriction for the "familiarity main effect" under the process of pooling variances, would be:

$$a_7 = a_8 = a_9.$$

The above restriction, substituted into Model 8-2, would yield:

(Model 8-5) $Y_1 = a_0U + a_{10}X_{10} + a_{11}X_{11} + E_5$

Model 8-5 has two linearly independent vectors: $df_1 = 4-2 = 2$ and $df_2 = 30-4 = 26$. The F value resulting from comparing Model 8-5 against Model 8-2 would be numerically equal to: 81.74/15.97.

Repeated Measures Design

Many of the problems presented in this book deal with experiments where subjects were randomly assigned to treatment groups and F tests were made to determine how likely the observed group differences were due to sampling error. On the other hand, experimental research in the behavioral sciences often is concerned with the performance of a few individuals who are examined over a number of experimental conditions. The N individuals may be trained in a specific task, and then the same N individuals may be required to perform the task under k conditions.

If one assumes the N*k observations are independent (uncorrelated), then the simple regression equation $\tilde{Y}_1 = a_0U + a_1X_1 \ . \ . \ . \ aX_k$ can be compared with the restricted equation $\tilde{Y}_1 = a_0U$ to determine the significance of task performance differences due to the k conditions. Note that Y_1 represents the task performance under k conditions, which means that the Y_1 vector will have N*k elements, rather than just N elements.

Under normal repeated measures conditions, one cannot assume that the N*k observations are independent because an individual who scores high in relation to the other subjects under one experimental condition tends to score high (in relation to the others) in subsequent treatment conditions. Likewise, the low-scoring subjects will tend to remain relatively low on subsequent performance. If the variance associated with these individual differences is ignored, the unaccounted for variance will be artificially large and the risk of making a Type II error is increased. One can overcome this problem by way of including a set of person vectors in the regression equation.

This might be best illustrated using an example provided by Winer (1962, page 112). Five subjects were randomly selected and trained on a reaction time task. The subjects were then required to perform under four different drug conditions with adequate time between conditions to rule out the combined effect of drugs. The sequence of drug administration was randomized; of course, counterbalancing of sequence could have been used. In the study there are twenty observations, four for each subject. If the investigator ignores the variance due to individual differences, Equation 8.4

$$Y_1 = a_0U + a_1X_1 + a_2X_2 + a_3X_3 + a_4X_4 + E_1 \quad (8.4)$$

can be cast to estimate the per cent of total reaction time variance due to drug conditions.

where:

X_1 = 1 if the score on the criterion was made by a person under drug condition 1; zero otherwise.

X_2 = 1 if the score on the criterion was made by a person under drug condition 2; zero otherwise.

X_3 = 1 if the score on the criterion was made by a person under drug condition 3; zero otherwise.

X_4 = 1 if the score on the criterion was made by a person under drug condition 4; zero otherwise.

Y_1 = reaction time.

The restriction $a_1 = a_2 = a_3 = a_4 = a_0$ can be made, and an F test made to determine how likely the observed mean drug differences were due to sampling error.

With the data provided by Winer (1962, page 112), the R^2 using the unrestricted equation was .468. There are four independent vectors in Equation 8.4 and one independent vector (the unit vector) in the restricted equation. Therefore, the F test will be:

$$F = \frac{(.468-0) \ / \ (4-1)}{(1-.468) \ / \ (20-4)} = 4.69$$

An F of 4.69 with 3 and 16 degrees of freedom occurs 1.5 times in a hundred. If the pre-determined alpha level was set at .01, then the hypothesis that $a_1 = a_2 = a_3 = a_4 = a_0$ would be accepted.

Since knowledge of individual differences are available, one can extract from the error variance the variance due to individual variability. Equation 8.5 reflects the unrestricted equation which uses all the experimental information available.

$$(8.5) \quad Y_1 = a_0U + a_1X_1 + a_2X_2 + a_3X_3 + a_4X_4 + a_5P_1 + a_6P_2 + a_7P_3 + a_8P_4 + a_9P_5 + E_2$$

where:

X_1 = 1 if the score on the criterion was made by a person under drug condition 1; zero otherwise.

X_2 = 1 if the score on the criterion was made by a person under drug condition 2; zero otherwise.

X_3 = 1 if the score on the criterion was made by a person under drug condition 3; zero otherwise.

X_4 = 1 if the score on the criterion was made by a person under drug condition 4; zero otherwise.

P_1 = 1 if the score on the criterion was made by person #1; zero otherwise.

P_2 = 1 if the score on the criterion was made by person #2; zero otherwise.

P_3 = 1 if the score on the criterion was made by person #3; zero otherwise.

P_4 = 1 if the score on the criterion was made by person #4; zero otherwise.

P_5 = 1 if the score on the criterion was made by person #5; zero otherwise.

Y_1 = reaction time.

These vectors with the criterion scores are presented below:

Y_1		U		X_1	X_2	X_3	X_4	P_1	P_2	P_3	P_4	P_5	E_2
30		1		1	0	0	0	1	0	0	0	0	
28		1		0	1	0	0	1	0	0	0	0	
16		1		0	0	1	0	1	0	0	0	0	
34		1		0	0	0	1	1	0	0	0	0	
14		1		1	0	0	0	0	1	0	0	0	
18		1		0	1	0	0	0	1	0	0	0	
10		1		0	0	1	0	0	1	0	0	0	
22		1		0	0	0	1	0	1	0	0	0	
24		1		1	0	0	0	0	0	1	0	0	
20	$= a_0$	1	$+ a_1$	0	1	0	0	0	0	1	0	0	
18		1		0	0	1	0	0	0	1	0	0	
30		1		0	0	0	1	0	0	1	0	0	
38		1		1	0	0	0	0	0	0	1	0	
34		1		0	1	0	0	0	0	0	1	0	
20		1		0	0	1	0	0	0	0	1	0	
44		1		0	0	0	1	0	0	0	1	0	
26		1		1	0	0	0	0	0	0	0	1	
28		1		0	1	0	0	0	0	0	0	1	
14		1		0	0	1	0	0	0	0	0	1	
30		1		0	0	0	1	0	0	0	0	1	

The first element in the Y_1 vector is 30 and the "1's" in vectors X_1 and P_1 indicate that the score was achieved by person #1 under drug condition 1. The second element (28) was also achieved by person #1 ($P_1 = 1$) but under drug condition 2 ($X_2 = 1$). The last element in vector Y_1 (30) represents the score of individual 5 ($P_5 = 1$) under drug condition 4 ($X_4 = 1$).

Equation 8.5, which includes the five person vectors, extracts the variance due to individual differences (between people). The RSQ using Equation 8.5 is <u>.9244</u>. Now if one hypothesizes $a_1 = a_2 = a_3 = a_4 = a_0$, the restricted model would be:

$\widetilde{Y}_1 = a_0U + a_5P_1 + a_6P_2 + a_7P_3 + a_8P_4 + a_9P_5$. The RSQ using the restricted model is <u>.4564</u>.

There are eight linearly independent vectors in the full model and five linearly independent vectors in the restricted model; therefore, the F ratio is:

$$F = \frac{(.9244 - .4564) / (8-5)}{(1 - .9244) / (20-8)} = \frac{.468/3}{.0756/12} = 24.7583$$

An F of 24.7583 with 3 and 12 degrees of freedom is observed 2 times in 10,000 ($p < .00002$). If the pre-determined alpha level was set at .01, then the hypothesis that $a_1 = a_2 = a_3 = a_4 = a_0$ is <u>rejected</u>. If you examine the two F tests, you should note that the variance due to the drug conditions is identical (.468); however, the error term is greatly reduced in the second F test due to the extraction of between-person variance.

Essentially, the person vector permits an adjustment of the predicted scores based on the average of a person's four scores.

Of course, any combination of drug conditions can be equated (e.g., $a_1 = a_2 \neq a_3 \neq a_4$) and tested using procedures outlined earlier in this book.

Since one usually finds individual differences in repeated measures studies, the procedures just outlined will provide a more sensitive test than analyses which ignore between-persons variance. Furthermore, persons vectors can be added to the complex models discussed earlier. The careful reader might note that in the special case where two repeated measures are made, this analysis provides the same probability statement one obtains with the correlated "t."

With respect to the LINEAR program, one would need to punch a subject card for each trial.

Person vectors can be generated by focusing on the subject identification, especially if the subjects have been assigned consecutive code numbers. For example, in the four-drug experiment given above, column 1 may be reserved for subject identification (1, 2, 3, 4, 5). Columns 2 and 3 may be the reaction time and column 4 a "1" if the score on Y_1 was observed under drug condition 1, zero otherwise, ... column 7 a "1" if the score on Y_1 was observed under drug condition 4, zero otherwise. The format would be (F1.0,F2.0,4F1.0) and this represents six original vectors. Three statements can be written for DATRAN which will generate the five person vectors:

```
K=A(I+1)
L=6+K+J
A(L)=1.0
```

The statement K=A(I+1) specifies K as the identification number. The machine reads the first card (I.D. #1) and since A(I+1) = 1,

K = 1 and L = 6+1+J. A new vector A(L)=1.0 is generated and because L=7, the new vector will be A(J+7). A(J+7) then will contain a "1" if the score on the criterion was made by individual #1; zero otherwise. If the first four cards represent individual #1, then the first four elements in A(J+7) will be "1's." The process will continue and produce new vectors for all values of K. In this case K will be 1 through 5, and five new vectors, A(J+7), A(J+8). . . A(J+11), will be formed. Note that the value 6 in L = 6+1+J represents the number of original vectors. If ten vectors were initially read in, then the value in L should be 10 (L = 10+1+J).

Mathematical Proof That r^2 Is Equal to the Variance of the Predicted Scores Divided by the Variance of the Actual Scores

In the area of prediction, one is constantly concerned with the accuracy of the prediction. One of the techniques for indicating the accuracy of a prediction equation is to determine the ratio of the variances of the predicted scores and the observed scores on the experienced sample (the sample from which the prediction equation was developed). The following shows that this ratio is actually r^2:

$$Y = bX + C$$

is a general equation for a straight line.

where:

Y = dependent variable

X = independent variable

b = slope of line

C = Y-intercept of line

Let $\tilde{Y}$ = predicted score; then $\tilde{Y} = bX + C$,

but $C = \bar{Y} - b\bar{X}$

$\therefore$ $\tilde{Y} = bX + \bar{Y} - b\bar{X}$

$$\tilde{Y} = b(X - \bar{X}) + \bar{Y}$$

$$\tilde{Y} = b(x) + \bar{Y}$$

but $(\tilde{Y} - \bar{Y}) = \tilde{y}$

$$\tilde{y} = b(x)$$

We also know that $b = \frac{\Sigma x_i y_i}{\Sigma x_i^2}$

$$b = \frac{\Sigma x_i y_i}{\sqrt{\Sigma x_i^2}\ \sqrt{\Sigma x_i^2}}$$

$$b = \frac{\Sigma x_i y_i}{\sqrt{\Sigma x_i^2}\ \sqrt{\Sigma x_i^2}} * \frac{\sqrt{\Sigma y_i^2}}{\sqrt{\Sigma y_i^2}}$$

$$b = \frac{\Sigma x_i y_i}{\sqrt{\Sigma x_i^2}\ \sqrt{\Sigma y_i^2}} * \frac{\sqrt{\Sigma y_i^2}}{\sqrt{\Sigma x_i^2}}$$

but $r = \frac{\Sigma x_i y_i}{\sqrt{\Sigma x_i^2}\ \sqrt{\Sigma y_i^2}}$

$$b = r\sqrt{\frac{\Sigma y_i^2}{\Sigma x_i^2}}$$

$$b = r\frac{\dfrac{\sqrt{\Sigma y_i^2}}{\sqrt{N}}}{\dfrac{\sqrt{\Sigma x_i^2}}{\sqrt{N}}} = r\frac{\sqrt{\dfrac{\Sigma y_i^2}{N}}}{\sqrt{\dfrac{\Sigma x_i^2}{N}}} = r\ \frac{S_y}{S_x}$$

Now let us use this information in our prediction equation:

$$\tilde{y} = b(x)$$

$$\tilde{y} = r\left(\frac{s_Y}{s_X}\right) X$$

Square the deviation of the predicted score:

$$\tilde{y}^2 = r^2 \frac{s_Y^2}{s_X^2} (x^2)$$

Sum the deviation scores:

$$\Sigma \tilde{y}_i^2 = r^2 \frac{s_Y^2}{s_X^2} \Sigma x_i^2$$

The variance of the predicted scores is:

$$\frac{\Sigma \tilde{y}_i^2}{N}$$

The variance of the observed scores is:

$$\frac{\Sigma y_i^2}{N}$$

The index of accuracy is:

$$\frac{\frac{\Sigma \tilde{y}_i^2}{N}}{\frac{\Sigma y_i^2}{N}} = \frac{\Sigma \tilde{y}_i^2}{\Sigma y_i^2}$$

But, as we have shown:

$$\Sigma \tilde{y}_i^2 = r^2 \frac{s_Y^2}{s_X^2} \Sigma x_i^2$$

Then the index of accuracy is the ratio of the variance of the predicted and observed scores.

$$\frac{r^2 \frac{s_Y^2}{s_X^2} \Sigma x_i^2}{\Sigma y_i^2}$$

$$\cfrac{\cfrac{\cfrac{r^2 \, \Sigma y_i^2 \, \Sigma x_i^2}{N}}{\cfrac{\Sigma x_i^2}{N}}}{\Sigma y_i^2}$$

$$\frac{\dfrac{r^2 \, \Sigma y_i^2 \, \Sigma x_i^2}{N} \qquad \dfrac{N}{\Sigma x_i^2}}{\Sigma y_i^2}$$

$$\frac{r^2 \, \Sigma y_i^2}{\Sigma y_i^2} = r^2$$

Therefore, the ratio of the variances of the predicted and observed scores is r^2.

The Mathematical Equivalence of the Tests of Significance of the Point Biserial Correlation Coefficient and the t-Test for the Difference between Means

In order for one to do effective research, there must be a basic research question that should be answered. Once that question has been stated, it is important to determine the statistical hypothesis that will help the researcher answer his question. Quite often researchers disagree with respect to the statistical technique to be employed. Usually much of this disagreement is a function of the training of the researcher and not a function of the statistics.

A good example of the different approaches is the situation in which the researcher is concerned with a comparison of two things (i.e., two instructional techniques in teaching undergraduate educational psychology). One researcher might want to compare the means of the two techniques while another might want to determine the correlation (point biserial correlation) between group membership and some continuous criterion. The following discussion shows the mathematical equivalence of the two techniques employed in hypothesis testing.

When a researcher is investigating group differences on a criterion, he is essentially wanting to know if the entities in one group exhibit a different level of the attribute than do the entities in the other group. The conventional approach is to investigate group differences in terms of mean difference, i.e., in terms of the analysis of variance model. If the design employs only two groups, the technique is termed a "t-test" analysis.

If a research question is couched in terms of the correlational model, the researcher is asking if there is a relationship between the two variables. For the purposes of this presentation, we will utilize only the correlational model which assumes a linear relationship. If there is a positive correlation between the two variables, the entities which have high scores on the one variable will also tend to have high scores on the other variable. Likewise, if there is a negative correlation, the entities which have high scores on the one variable will tend to have low scores on the other variable.

Now consider the point biserial correlational model with the dichotomous attribute indicating group membership (a "1" for Group A, a "2" for Group B, for instance). With this latter model we are asking if there is a relationship between group membership and the continuous attribute (criterion). If there is a positive correlation, then we know that the higher an entity's group membership score, the higher the score on the criterion. If there is a negative correlation, those entities which have a high group membership score will tend to have lower scores on the criterion. Because of the positive correlation for the data in Table 8-2, we can ascertain that the entities in Group A have a higher mean value on the criterion than those entities in Group B.

The correlational research question can be stated as follows: "Can we differentially predict the dependent attribute on the basis of knowledge of group membership?" It should be noted that we can differentially predict the dependent attribute only if there are differences between group means. Thus, we have arrived at the analysis of variance ("t-test," for the restricted two-group case) question again: "Are there differences (in the mean value of the criterion) between the two groups, or are there differences between group means?"

The preceding discussion has attempted to indicate that the researcher is asking basically the same question when he utilizes either of the following models: (a) the analysis of variance model for investigating the differences between two means, or (b) the correlation model for investigating the differential predictability of a dichotomous variable. The burden of the next section will be to show the mathematical equivalence of the tests of significance for these two models. Following the proof, we will present the linear regression model which encompasses both of the previously mentioned models.

There are many formulae available for the following statistics. Judicious choice of formulae would make the burden of the proof easier, except that the following proof has been found to be most easily understood by the student. In the following formulae, subscripts are used to identify Groups 1 and 2.

The test of significance for the difference between two uncorrelated means for this particular hypothesis ($\mu_1 - \mu_2 = 0$) is:

$$(1) \qquad t = \frac{(\overline{Y}_1 - \overline{Y}_2) - (\mu_1 - \mu_2)}{S_{\overline{Y}_1 - \overline{Y}_2}} = \frac{\overline{Y}_1 - \overline{Y}_2}{S_{\overline{Y}_1 - \overline{Y}_2}}$$

with degrees of freedom equal to $N_1 + N_2 - 2$. The formula for the unbiased estimate of the standard error of the difference between means is given by Runyon and Haber (1967, page 160) as:

$$(2) \qquad S_{\overline{Y}_1 - \overline{Y}_2} = \sqrt{\frac{(\Sigma Y_1{}^2 + \Sigma Y_2{}^2) - (N_1 \overline{Y}_1{}^2 + N_2 \overline{Y}_2{}^2)}{(N_1 + N_2 - 2)} \left(\frac{1}{N_1} + \frac{1}{N_2}\right)}$$

The test of significance for a correlation coefficient for this particular hypothesis ($\rho = 0$) is:

$$(3) \qquad t = r \frac{\sqrt{N-2}}{\sqrt{1-r^2}} = r\sqrt{\frac{N-2}{1-r^2}}$$

with degrees of freedom equal to N-2. Since $N_1 + N_2 - 2$ is equal to N-2, the degrees of freedom for both t-tests are the same. Our only problem now is to establish the mathematical equivalence of (1) and (3). The proof begins with (3) and ends with the expression for (1).

Considering the data in Table 8-2, we can determine the following identities:

$$(4) \qquad \Sigma X = N_1$$

$$(5) \qquad \overline{X} = N_1/N$$

Group	Dichotomous Attribute (X)	Continuous Attribute (Y)	$(Y)^2$
A	1	9	81
	1	11	121
	1	9	81
	1	11	121
B	0	6	36
	0	6	36
	0	7	49
	0	7	49
	0	8	64
	0	8	64

$\Sigma X = N_1 = 4$ $\qquad \bar{X} = \frac{N_1}{N} = \frac{4}{10}$

$\Sigma Y_1 = 40$

$\Sigma Y_2 = 42$ $\qquad \bar{Y}_1 = \frac{40}{4} = 10$

$\Sigma Y_1^2 = 404$

$\Sigma Y_2^2 = 298$ $\qquad \bar{Y}_2 = \frac{42}{6} = 7$

$$t = \frac{(\bar{Y}_1 - \bar{Y}_2) - (\mu_1 - \mu_2)}{\sqrt{\frac{(\Sigma Y_1^2 + \Sigma Y_2^2) - (N_1\bar{Y}_1^2 + N_2\bar{Y}_2^2)}{N_1 + N_2 - 2}\left(\frac{1}{N_1} + \frac{1}{N_2}\right)}} \approx 4.65$$

$$r = \frac{\Sigma XY - \frac{(\Sigma X)(\Sigma Y)}{N}}{\sqrt{\left(\Sigma X^2 - \frac{(\Sigma X)^2}{N}\right)\left(\Sigma Y^2 - \frac{(\Sigma Y)^2}{N}\right)}} \approx .854$$

$$t = r_{pbi}\sqrt{\frac{(N - 2)}{(1 - r_{pbi}^2)}} \approx 4.64$$

Table 8-2. Numerical example of the equivalence of the tests of significance for $\rho_{pbi} = 0$ and $\mu_1 = \mu_2$.

(6) $\Sigma X^2 = N_1$

(7) $\Sigma Y = \Sigma Y_1 + \Sigma Y_2$

(8) $\Sigma Y^2 = \Sigma {Y_1}^2 + \Sigma {Y_2}^2$

(9) $(\Sigma Y)^2 = (\Sigma Y_1 + \Sigma Y_2)^2 = (N_1\overline{Y}_1 + N_2\overline{Y}_2)^2 = {N_1}^2{\overline{Y}_1}^2 + {N_2}^2{\overline{Y}_2}^2 + 2N_1N_2\overline{Y}_1\overline{Y}_2$

(10) $\Sigma XY = \Sigma X_1Y_1 + \Sigma X_2Y_2 = \Sigma X_1Y_1 + 0 = \Sigma Y_1$

(11) $\overline{Y} = \dfrac{N_1\overline{Y}_1 + N_2\overline{Y}_2}{N_1 + N_2}$

(12) $\overline{Y}_1 - \overline{Y} = \overline{Y}_1 - \left(\dfrac{N_1\overline{Y}_1 + N_2\overline{Y}_2}{N_1 + N_2}\right) = \dfrac{N_2\,(\overline{Y}_1 - \overline{Y}_2)}{N_1 + N_2}$

(13) $\overline{Y}_1 - \overline{Y}_2 = \dfrac{N}{N_2}\,(\overline{Y}_1 - \overline{Y})$

We shall first express the traditional correlation coefficient in terms of the above relationships. Given the raw score formula for correlation:

(14)

$$r = \frac{\Sigma XY - \dfrac{\Sigma X\,\Sigma Y}{N}}{\sqrt{\left(\Sigma X^2 - \dfrac{(\Sigma X)^2}{N}\right)\left(\Sigma Y^2 - \dfrac{(\Sigma Y)^2}{N}\right)}}$$

We can substitute identities (4) and (10) to arrive at:

(15)

$$r = \frac{\Sigma Y_1 - N_1\dfrac{\Sigma Y}{N}}{\sqrt{\left(N_1 - \dfrac{{N_1}^2}{N}\right)\left(\Sigma Y^2 - \dfrac{(\Sigma Y)^2}{N}\right)}}$$

Now, by multiplying the numerator and denominator by $\frac{1}{N_1}$, we get:

(16)

$$r = \frac{\dfrac{1}{N_1}\left[\Sigma Y_1 - N_1\dfrac{\Sigma Y}{N}\right]}{\sqrt{\dfrac{1^2}{{N_1}^2}\left[\left(N_1 - \dfrac{{N_1}^2}{N}\right)\left(\Sigma Y^2 - \dfrac{(\Sigma Y)^2}{N}\right)\right]}},$$

which can be simplified by collecting terms in the numerator and by use of a common denominator for some of the elements within the radical:

(17)
$$r = \frac{\frac{\Sigma Y_1}{N_1} - \frac{\Sigma Y}{N}}{\sqrt{\frac{1}{N_1{}^2}\left(\frac{NN_1 - N_1{}^2}{N}\right)\left(\Sigma Y^2 - \frac{\Sigma Y^2}{N}\right)}}$$

If we now utilize our definition of the mean and factor the second element under the radical, we can simplify (17) to read:

(18)
$$r = \frac{\bar{Y}_1 - \bar{Y}}{\sqrt{\left(\frac{N_1(N-N_1)}{N_1{}^2 N}\right)\left(\Sigma Y^2 - \frac{(\Sigma Y)^2}{N}\right)}}$$

Now by using the relationship that $N_1 + N_2 = N$ and simplifying $N_1/N_1{}^2$ to $1/N_1$, we get:

(19)
$$r = \frac{\bar{Y}_1 - \bar{Y}}{\sqrt{\left(\frac{N_2}{N_1}\right)\left(\frac{1}{N}\right)\left(\Sigma Y^2 - \frac{(\Sigma Y)^2}{N}\right)}}$$

By collecting terms we arrive at:

(20)
$$r = \frac{\bar{Y}_1 - \bar{Y}}{\sqrt{\left(\frac{N_2}{N_1}\right)\left(\frac{\Sigma Y^2}{N} - \frac{(\Sigma Y)^2}{N}\right)}}$$

and by utilizing our definition of variance, we arrive at our final expression for the correlation coefficient when dealing with the point biserial model:

(21)
$$r_{pbi} = \frac{\overline{Y}_1 - \overline{Y}}{\sqrt{\left(\frac{N_2}{N_1}\right) \quad S_Y^2}}$$

Now let us replace this formulae for r into the test of significance (3). Repeating the test of significance (3) for correlation,

(22)
$$t = r\sqrt{\frac{N - 2}{1 - r^2}}$$

and replacing the formula (21) for r into (22), we get:

(23)
$$t = \left(\frac{\overline{Y}_1 - \overline{Y}}{\sqrt{\frac{N_2}{N_1} \quad S_Y^2}}\right)\sqrt{\frac{N - 2}{1 - \frac{(\overline{Y}_1 - \overline{Y})^2}{\frac{N_2}{N_1} \quad S_Y^2}}}$$

If we now collect all terms under the radical,

(24)
$$t = (\overline{Y}_1 - \overline{Y})\sqrt{\frac{N - 2}{\left(\frac{N_2}{N_1} \quad S_Y^2\right) - (\overline{Y}_1 - \overline{Y})^2}}$$

and if we multiply numerator and denominator by N/N_2, we now have:

(25)
$$t = \frac{N}{N_2}(\overline{Y}_1 - \overline{Y})\sqrt{\frac{N - 2}{\left(\frac{N_2}{N_1} \quad S_Y^2\right) - (\overline{Y}_1 - \overline{Y})^2}\left(\frac{N_2^2}{N^2}\right)}$$

By (13) and inverting the radical, we have:

(26)
$$t = \frac{\overline{Y}_1 - \overline{Y}_2}{\sqrt{\frac{\left(\frac{N_2}{N_1} \quad S_Y^2\right) - (\overline{Y}_1 - \overline{Y})^2}{N - 2}\left(\frac{N^2}{N_2^2}\right)}}$$

We note that the numerator is equal to that of (3), so our task now is to simplify the expression under the radical. We can factor out N/N_1N_2 and get the following expression:

(27)

$$t = \frac{\bar{Y}_1 - \bar{Y}_2}{\sqrt{\dfrac{\left[S_Y{}^2 - \dfrac{N_1}{N_2} - \left((\bar{Y}_1 - \bar{Y})^2\right]N\right.}{N - 2}\left(\dfrac{N}{N_1 N_2}\right)}}$$

Now substitute (12) into the above equation:

(28)

$$t = \frac{\bar{Y}_1 - \bar{Y}_2}{\sqrt{\dfrac{\left[S_Y{}^2 - \dfrac{N_1}{N_2}\left(\left(\dfrac{N_2}{N_1}\right)(\bar{Y}_1 - \bar{Y}_2)\right)^2\right]N}{N - 2}\left(\dfrac{N}{N_1 N_2}\right)}}$$

and by definition of variance and carrying out the squaring operation, we get:

(29)

$$t = \frac{\bar{Y}_1 - \bar{Y}_2}{\sqrt{\dfrac{\left(\left[\dfrac{\Sigma Y^2}{N} - \dfrac{(\Sigma Y)^2}{N^2}\right] - \dfrac{N_1}{N_2}\left[\dfrac{N_2{}^2(\bar{Y}_1{}^2 + \bar{Y}_2{}^2 - 2\bar{Y}_1\bar{Y}_2)}{N_1{}^2}\right]\right)N}{N - 2}\left(\dfrac{N}{N_1 N_2}\right)}}$$

Now we can simplify (29) into:

(30)

$$t = \frac{\bar{Y}_1 - \bar{Y}_2}{\sqrt{\dfrac{\left(\Sigma Y^2 - \dfrac{(\Sigma Y)^2}{N}\right) - \dfrac{N_1 N_2}{N}(\bar{Y}_1{}^2 + \bar{Y}_2{}^2 - 2\bar{Y}_1\bar{Y}_2)}{N - 2}\left(\dfrac{N}{N_1 N_2}\right)}}$$

and then by utilizing (9), we have:

(31)

$$t = \frac{\bar{Y}_1 - \bar{Y}_2}{\sqrt{\dfrac{\dfrac{\Sigma Y^2 - (N_1{}^2\bar{Y}_1{}^2 + N_2{}^2\bar{Y}_2{}^2 + 2N_1N_2\bar{Y}_1\bar{Y}_2)}{N} - \dfrac{N_1N_2}{N}(Y_1{}^2 + Y_2{}^2 - 2Y_1Y_2)}{N - 2}\left(\dfrac{N}{N_1 N_2}\right)}}$$

Now by expansion we have:

(32)

$$t = \frac{\bar{Y}_1 - \bar{Y}_2}{\sqrt{\dfrac{\left[\dfrac{\Sigma Y^2 - N_1{}^2Y_1{}^2}{N} - \dfrac{N_2\bar{Y}_2{}^2}{N} - \dfrac{2N_1N_2\bar{Y}_1\bar{Y}_2}{N} - \dfrac{N_1N_2\bar{Y}_1{}^2}{N} - \dfrac{N_1N_2\bar{Y}_2{}^2}{N} + \dfrac{2N_1N_2Y_1Y_2}{N}\right]}{N - 2}\left(\dfrac{N}{N_1 N_2}\right)}}$$

Now we can factor some of these elements under the radical:

(33)
$$t = \frac{\bar{Y}_1 - \bar{Y}_2}{\sqrt{\dfrac{\Sigma Y^2 - \left[\dfrac{(N_1+N_2)}{N} (N_1\bar{Y}_1{}^2 + N_2\bar{Y}_2{}^2)\right]}{N - 2} \left(\dfrac{N}{N_1N_2}\right)}}$$

and this equation can be further simplified, because $N_1 + N_2 = N$, and we can also rewrite (N/N_1N_2) as: $(\frac{1}{N_1} + \frac{1}{N_2})$; therefore:

(34)
$$t = \frac{\bar{Y}_1 - \bar{Y}_2}{\sqrt{\dfrac{\Sigma Y^2 - (N_1\bar{Y}_1{}^2 + N_2\bar{Y}_2{}^2)}{N - 2} \left(\dfrac{1}{N_1} + \dfrac{1}{N_2}\right)}}$$

but replacing the identity (8) into this last expression gives us:

(35)
$$t = \frac{\bar{Y}_1 - \bar{Y}_2}{\sqrt{\dfrac{(\Sigma Y_1{}^2 + \Sigma Y_2{}^2) - (N_1\bar{Y}_1{}^2 + N_2\bar{Y}_2{}^2)}{N - 2} \left(\dfrac{1}{N_1} + \dfrac{1}{N_2}\right)}}$$

which is the formula for the t-test for the significance of the difference between two means under the specified hypothesis (1).

The linear regression analogue (Bottenberg and Ward, 1963) for testing the difference between two means is as follows. A "full" regression model is constructed which is defined as follows:

Model 1 $\quad Y_1 = a_0U + a_1X_1 + a_2X_2 + E_1$

where:

Y_1 = the criterion variable.

U = the unit vector.

X_1 = the group membership vector for Group 1 and contains a "1" if the individual is in Group 1, and a "0" if the individual is not a member of Group 1.

X_2 = the group membership vector for Group 2 and contains a "1" if the individual is in Group 2, and a "0" if the individual is not a member of Group 2.

E_1 = the error vector, or the discrepancy between the predicted Y_1 value and the actual Y_1 value.

a_0, a_1, and a_2 are unknown least squares weights.

The amount of variance that Model 1 accounts for is tested against the amount of variance that the following restriction explains. Under the null hypothesis of no group differences, $a_1 = a_2$ and the restricted model is thus:

Model 2 $\quad Y_1 = a_0U + E_2$

where:

Y_1 = the same dependent variable.

U = the same unit vector.

E_2 = a new error vector.

a_0 = another least squares weight.

Model 2 is considered a restricted model because we are not taking into account information about group membership, i.e., we have added together the two group membership vectors in accordance with our null hypothesis that $a_1 = a_2$, and we are predicting the dependent variable on the basis of the overall group mean. When we test the full model (Model 1) against the restricted model (Model 2), we are essentially asking if we can significantly improve our prediction of the dependent variable by taking into account group membership. Parenthetically, it should make sense that we can improve prediction of the dependent variable (when utilizing group membership) only if the two groups have somewhat different mean values on the dependent variable. If the two groups have the same mean value on the dependent variable, then the overall group mean will be the same as the mean of either group, and we cannot predict the dependent variable any better by knowing group membership (Model 1) than by simply knowing the overall group mean (Model 2).

Model 2 is simply the correlation between the dependent scores and the vector containing all "1's." Since there is no variability in the scores in the unit vector, this correlation must, of necessity, be zero.

Model 1 is the correlation between the dependent variable and the linear composite of three variables (U, X_1, and X_2), but these three variables can actually be represented by a single variable. We have already discussed the predictive efficiency of the unit vector. If we ignore this vector, we do not decrease the magnitude of our relationship.

X_1 and X_2 provide the same information, for if a "1" appears in X_1, we know that a "0" must appear in X_2 (i.e., if a person is a member of Group 1, he is not a member of Group 2 and vice versa). Therefore, we can eliminate either X_1 or X_2 from the linear composite, and we now have only one variable (either X_1 or X_2) being correlated with our dependent variable. The dependent variable is a continuous variable, but the group membership vector is, of necessity, a dichotomous vector. Model 1 has a specific name which we have previously discussed -- the point biserial correlation coefficient. Thus, when the question of significance is raised, the correlational model and the analysis of variance model should give us the same results as does the linear regression model.

The conventional F test for linear regression models (Bottenberg and Ward, 1963) is as follows:

$$F = \frac{(R_F{}^2 - R_R{}^2) / df_1}{(1 - R_F{}^2) / df_2} \quad (36)$$

where:

$R_F{}^2$ = the square of the correlation produced by the full model, or the amount of variance accounted for by the full model.

$R_R{}^2$ = the square of the correlation produced by the restricted model, or the amount of variance accounted for by the restricted model.

df_1 = the degrees of freedom for the numerator, computed by subtracting the number of linearly independent vectors in the restricted model from the number of linearly independent vectors in the full model.

df_2 = the degrees of freedom for the denominator, computed by subtracting the number of linearly independent vectors in the full model from the total number of subjects.

In the particular problem being discussed, the full model has two linearly independent vectors (the unit vector and one group membership vector), while the restricted model has only one linearly independent vector (the unit vector). Again, the correlation produced by the full model is simply the point biserial correlation (or r_{pbi}), and the correlation produced by the restricted model is zero. In symbolical terms:

(37) $R_F{}^2 = r_{pbi}{}^2$

(38) $R_R{}^2 = 0$

(39) $df_1 = 2-1 = 1$

(40) $df_2 = N-2$

Substituting these values into the above formula for F, we have:

(41)
$$F = \frac{(r_{pbi}^2 - 0) / 1}{(1 - r_{pbi}^2) / (N-2)} = \frac{r_{pbi}^2}{(1 - r_{pbi}^2) / (N-2)}$$

Now let us recall that $F = t^2$ when <u>one</u> degree of freedom is present in the numerator of the F statistic (Hays, 1963, page 354). Taking the square root of both sides of the above equation will give us an expression in terms of t:

(42)
$$t = \sqrt{F} = \sqrt{\frac{r_{pbi}^2}{(1 - r_{pbi}^2) / (N-2)}}$$

(43)
$$t = r_{pbi}\sqrt{\frac{1}{(1 - r_{pbi}^2) / (N-2)}}$$

(44)
$$t = r_{pbi}\sqrt{\frac{N - 2}{1 - r_{pbi}^2}}$$

The above equation is the conventional formula for testing the hypothesis that the population correlation is equal to zero and was derived from the linear regression approach in testing the hypothesis that the difference between two means is equal to zero.

The primary purpose of this discussion was to show that the t-test (analysis of variance) for the difference between two means ($\mu_1 - \mu_2 = 0$) is mathematically equivalent to the statistical test for the point biserial correlation coefficient ($\rho_{pbi} = 0$).

Having shown that these two tests of significance are equivalent, we can return to the basic research questions and emphasize that either statistical test will answer the stated research questions. With this in mind, it would seem that the researcher would concern himself more with the research question and realize that the comprehensive regression model could be used to answer the research question.

One final note of discord should be added. If the point biserial technique is used, the means of the criterion in each group are not readily available. Conversely, if the t-test (analysis of variance) is used, the relationship between group membership and the criterion is not known. With little effort in either technique, the additional information can be obtained, although with the regression model this information is readily available for interpretation.

Appendix A

Answers to Problem Set for Chapter V

1. Your output should look like Figures 5-5 and 5-6.

2. Your output should look like Figure 5-7.

3. (a) title card N=36 (or anything you wish to put)

   parameter card 000360001200012

   (b) title card N=180

   parameter card 001800002000020

   (c) title card N=1000

   parameter card 010000000800008

4. (a) MODEL 1 007001.E-0501010303050802

   (b) MODEL 2 005001.E-050809111205

   (c) MODEL 3 007001.E-0502020405111107

   (d) MODEL 4 005001.E-050202040507

   (e) MODEL 5 003001.E-05020207

5. (a) F RATIO 1 6 030400010032

   (b) F RATIO 2 6 030500020032

   (c) F RATIO 3 6 040500010033

6. title card REACTION TIME N=50

   parameter card 000500000400004

   format card (2X,2F1.0,2F2.0)

   model cards MODEL 1 003001.E-05010203

   MODEL 2 003001.E-05010204

   F Ratio cards F RATIO 1 6 019000010048

   F RATIO 2 6 029000010048

Answers to Problem Set for Chapter VI

Line Plotting Problem:

A. $\widetilde{Y}_1 = a_0U + a_4X_4 + a_5X_5$

1. If $X_{5_a} = 4$:

$$\widetilde{Y}_{1_a} = \left\{ \begin{array}{lll} a_0U & = 3.85*1 & = 3.85 \\ a_4X_{4_a} & = 3.12*1 & = 3.12 \\ a_5X_{5_a} & = 0.13*4 & = \underline{0.52} \\ & & 7.49 \end{array} \right.$$

2. If $X_{5_b} = 6$:

$$\widetilde{Y}_{1_b} = \left\{ \begin{array}{lll} a_0U & = 3.85*1 & = 3.85 \\ a_4X_{4_a} & = 3.12*1 & = 3.12 \\ a_5X_{5_b} & = 0.13*6 & = \underline{0.78} \\ & & 7.75 \end{array} \right.$$

B. $\widetilde{Y}_1 = a_0U + a_5X_5$

1. If $X_{5_c} = 2$:

$$\widetilde{Y}_{1_c} = \left\{ \begin{array}{lll} a_0U & = 4.95*1 & = 4.95 \\ a_5X_{5_c} & = 0.13*2 & = \underline{0.26} \\ & & 5.21 \end{array} \right.$$

2. If $X_{5_d} = 8$:

$$\widetilde{Y}_{1_d} = \left\{ \begin{array}{lll} a_0U & = 4.95*1 & = 4.95 \\ a_5X_{5_d} & = 0.13*8 & = \underline{1.04} \\ & & 5.99 \end{array} \right.$$

C. $\widetilde{Y}_1 = a_0U + a_4X_4 + a_8X_8$

1. If $X_{8_e} = 3$:

$$\widetilde{Y}_{1_e} = \left\{ \begin{array}{lll} a_0U & = -1.36*1 & = -1.36 \\ a_4X_{4_e} & = 12.23*1 & = 12.23 \\ a_8X_{8_e} & = -0.53*3 & = \underline{-1.59} \\ & & 9.28 \end{array} \right.$$

2. If $X_{8_f} = 8$:

$$\widetilde{Y}_{1_f} = \left\{ \begin{array}{lll} a_0U & = -1.36*1 & = -1.36 \\ a_4X_{4_f} & = 12.23*1 & = 12.23 \\ a_8X_{8_f} & = -0.53*8 & = \underline{-4.24} \\ & & 6.63 \end{array} \right.$$

Problem 6-1:

a. title card: Test for Group Membership Difference N=24 (may vary

parameter card: 000240000500005

format card: (2X,F2.0,3F1.0,F2.0)

b. $Y_1 = a_0U + a_2X_2 + a_3X_3 + a_4X_4 + a_5X_5 + E_1$

where:

X_2 = 1 if member of Condition 1; zero otherwise.

X_3 = 1 if member of Condition 2; zero otherwise.

X_4 = 1 if member of Condition 3; zero otherwise.

X_5 = test anxiety level

and where:

weights a_0, a_2, a_3, a_4, and a_5 are calculated to minimize the sum of the squared elements in vector E_1.

We have used these numbers to correspond to the order in which the variables are read by the computer and in which they will be shown on the model card. This is not at all necessary, as long as your explanation of your model is correct and you then write a model card that will reflect the same information.

c. $Y_1 = a_0U + a_5X_5 + E_2$

where: X_5 = test anxiety level

and where: a_0 and a_5 are calculated to minimize the sum of the squared elements in vector E_2.

d. MODEL 1 003001.E-05020501

MODEL 2 003001.E-05050501

e. F RATIO 1 6 010200020020

The degrees of freedom in Model 1 = 4.
The degrees of freedom in Model 2 = 2.

Problem 6-2:

a. $Y_1 = a_0U + a_1X_1 + a_2X_2 + a_3X_3 + a_4X_4 + a_5X_5 + a_6X_6 + E_1$

where: X_1, X_2, and X_3 = mutually exclusive group membership vectors.

and where: X_4, X_5, and X_6 = the appropriate anxiety score (e.g., X_4 = anxiety score if the score on Y_1 represents a member of group 1; zero otherwise).

b. $Y_1 = a_0U + a_1X_1 + a_2X_2 + a_3X_3 + a_7X_7 + E_2$

where: $X_7 = X_4 + X_5 + X_6$

The restriction is expressed as $a_4 = a_5 = a_6 = a_7$, a common weight. Since X_7 would be a vector containing the anxiety scores for all subjects, it is the same as our original vector for anxiety scores, read by the format card as vector 5.

c. title card: Test for Interaction N=24 (may vary)

parameter card: 000240000500008

format card: (2X,F2.0,3F1.0,F2.0)

d. MODEL 1 005001.E-050204060801

MODEL 2 003001.E-05020501

e. F RATIO 1 6 010200020018

Problem 6-3:

a. A(J+3)=A(I+2)**2

A(J+4)=A(I+2)**3

b. $Y_1 = a_0U + a_2X_2 + a_3X_2^{\ 2} + a_4X_2^{\ 3} + E_3$

c. $a_4 = 0$

$Y_1 = a_0U + a_2X_2 + a_3X_2^{\ 2} + E_4$

d. $a_3 = 0$

$$Y_1 = a_0U + a_2X_2 + E_5$$

Problem 6-5:

The full model is:

$$Y_1 = a_0U + a_1X_1 + a_2X_2 + a_3X_3 + a_4X_4 + a_5X_5 + a_6X_6 + E_1$$

where:

X_1 = 1 if the score on Y_1 belongs to a member of low-conforming group; zero otherwise (vector 2 for the computer).

X_2 = 1 if the score on Y_1 belongs to a member of high-conforming group; zero otherwise (vector 3 for the computer).

X_3 = authoritarian score, if score on Y_1 is low-conforming; zero otherwise (vector 5 for the computer).

X_4 = authoritarian score, if score on Y_1 is high-conforming; zero otherwise (vector 6 for the computer).

$X_5 = X_3^2$ (vector 7)

$X_6 = X_4^2$ (vector 8)

Transformation cards needed:

```
A(J+5)=A(I+2)*A(I+4)

A(J+6)=A(I+3)*A(I+4)

A(J+7)=A(J+5)**2

A(J+8)=A(J+6)**2

A(J+9)=A(J+7)+A(J+8)
```

The model, rewritten with vector subscripts:

$$Y_1 = a_0U + a_2X_2 + a_3X_3 + a_5X_5 + a_6X_6 + a_7X_7 + a_8X_8 + E_1 \text{(Model 1)}$$

Question 1: Is the RSQ of the full model probably zero?

$a_2 = a_3 = a_5 = a_6 = a_7 = a_8 = a_0$

$Y_1 = a_0U + E_2$

If no, Question 2.

Question 2: Is the influence of conformity group on leadership linear and different across the observed authoritarian scores?

$a_7 = a_8 = 0$

$Y_1 = a_0U + a_2X_2 + a_3X_3 + a_5X_5 + a_6X_6 + E_3$ (Model 2)

If yes (the loss of precision is not significant), ask Question 3. If no (it is significant), ask Question 5.

Question 3: Is the influence of conformity on leadership score constant across the observed authoritarian measure?

$a_5 = a_6 = a_4$ (a common weight)

$Y_1 = a_0U + a_2X_2 + a_3X_3 + a_4X_4 + E_4$ (Model 3)

If yes, ask Question 4. If no, plot the linear interaction, Model 2.

Question 4: Is the influence of conformity group on leadership equal (or the same) across the observed authoritarian scores?

$a_2 = a_3 = a_0$

$Y_1 = a_0U + a_4X_4 + E_5$ (Model 4)

If yes, simple linear relationship exists. If no, plot parallel lines, Model 3.

Question 5: (If the answer to Question 2 was "No") Is the non-linear influence of conformity group membership on leadership constant across the observed authoritarian scores?

$a_5 = a_6 = a_4$ (a common weight) and

$a_7 = a_8 = a_9$ (a common weight)

$Y_1 = a_0U + a_2X_2 + a_3X_3 + a_4X_4 + a_9X_9 + E_6$ (Model 5)

(Compare to Model 1) If yes, ask Question 6. If no, plot curved interaction, Model 4.

Question 6: Is the non-linear influence of conformity group membership on leadership equal across the observed authoritarian scores?

$a_2 = a_3 = a_0$

$Y_1 = a_0U + a_4X_4 + a_9X_9 + E_7$ (Model 6)

If yes, simple curved relationship exists. If no, plot parallel curves, Model 5.

We could run these questions in the computer one at a time and stop when we get the answer. But it will be easier to ask all of the questions in one run and then examine the output for the correct answer. It may be that this way we will ask questions that we won't need answered. (For example, if the answer to Question 1 is "No," to Question 2 is "Yes," and to Question 3 is "No," we then know we have a linear interaction and can stop.) But we can always ignore the extra output if it turns out to be extra. And this way we will avoid re-running the data.

PROBLEM 6-1

CORRELATIONS

	1	2	3	4	5
1	1.0000	-0.2837	-0.3207	0.6043	0.1197
2	-0.2837	1.0000	-0.5000	-0.5000	0.0000
3	-0.3207	-0.5000	1.0000	-0.5000	0.0000
4	0.6043	-0.5000	-0.5000	1.0000	0.0000
5	0.1197	0.0000	0.0000	0.0000	1.0000

```
.....MODEL 1      0.000010
 CRITERION     1
PREDICTORS      2 -   5

   RSQ = 0.38000515                                  4
```

VAR. NUMBER	STD. WT.	ERROR
2	0.02437941	-0.00021999
3	0.00000000	0.00021129
4	0.61653825	0.00000869
5	0.11967552	-0.00000000

VAR. NUMBER	WEIGHT
2	0.12354298
3	0.00000000
4	3.12431556
5	0.12968917

CONSTANT= 3.86378995

```
.....MODEL 2      0.000010
 CRITERION     1
PREDICTORS      5 -   5

   RSQ = 0.01432223                                  2
```

VAR. NUMBER	STD. WT.	ERROR

```
      5     0.23935104     0.11967552


 VAR. NUMBER      WEIGHT

       5        0.25937834
CONSTANT=       4.18448555

 ****F-RATIO =    5.8982    D.F. NUM. =   2. D.F. DEN. =  20.      1 0.3800   2 0.0143   PROB = 0.00969

TIME FOR THIS RUN IS       5.60 SECONDS

END-OF-DATA ENCOUNTERED ON SYSTEM INPUT FILE.
```

PROBLEM 6-2
MEANS-STANDARD DEVIATIONS-CORRELATIONS

MEANS

	1	2	3	4	5	6	7	8
1	5.7083	0.3333	0.3333	0.3333	5.8750	1.9583	1.9583	1.9583

STANDARD DEVIATIONS

	1	2	3	4	5	6	7	8
1	2.3888	0.4714	0.4714	0.4714	2.2044	3.0479	3.0479	3.0479

CORRELATIONS

	1	2	3	4	5	6	7	8
1	1.0000	-0.2837	-0.3207	0.6043	0.1197	-0.0246	-0.3221	0.4332
2	-0.2837	1.0000	-0.5000	-0.5000	0.0000	0.9086	-0.4543	-0.4543
3	-0.3207	-0.5000	1.0000	-0.5000	0.0000	-0.4543	0.9086	-0.4543
4	0.6043	-0.5000	-0.5000	1.0000	0.0000	-0.4543	-0.4543	0.9086
5	0.1197	0.0000	0.0000	0.0000	1.0000	0.2411	0.2411	0.2411
6	-0.0246	0.9086	-0.4543	-0.4543	0.2411	1.0000	-0.4128	-0.4128
7	-0.3221	-0.4543	0.9086	-0.4543	0.2411	-0.4128	1.0000	-0.4128
8	0.4332	-0.4543	-0.4543	0.9086	0.2411	-0.4128	-0.4128	1.0000

..... MODEL 1 0.000010
CRITERION 1
PREDICTORS 2 - 4
6 - 8

RSQ = 0.75993962 56

VAR. NUMBER	STD. WT.	ERROR
2	0.00000000	0.00163785
3	1.33153020	-0.00163800
4	2.41759741	0.00000015
6	1.33176637	0.00049874
7	-0.16318214	0.00081822
8	-0.67856556	-0.00243177

VAR. NUMBER	WEIGHT
2	0.00000000
3	6.74754655
4	12.25120628
6	1.04378349
7	-0.12789542
8	-0.53183167

CONSTANT=. -1.37669444

..... MODEL 2 0.000010
CRITERION 1
PREDICTORS 2 - 5

RSQ = 0.38000515

VAR. NUMBER	STD. WT.	ERROR
2	0.02437941	-0.00021999
3	0.00000000	0.00021129
4	0.[illegible]1653825	0.00000369
5	0.[illegible]1967552	-0.00000000

VAR. NUMBER	WEIGHT
2	0.12354298
3	0.00000000
4	3.12431556
5	0.12968917

CONSTANT= 3.86378995

****F-RATIO = 14.2440 D.F. NUM. = 2. D.F. DEN. = 18. 1 0.7599 2 0.3800 PROB = 0.00020

TIME FOR THIS RUN IS 5.97 SECONDS

END-OF-DATA ENCOUNTERED ON SYSTEM INPUT FILE.

```
PROBLEM 6-3
   CORRELATIONS

                 1            2            3            4

       1      1.0000       0.8351       0.8697       0.8895
       2      0.8351       1.0000       0.9795       0.9365
       3      0.8697       0.9795       1.0000       0.9873
       4      0.8895       0.9365       0.9873       1.0000


.....MODEL 1      0.000010
 CRITERION     1
PREDICTORS      2 -   4


   RSQ = 0.92832471                                          16


VAR. NUMBER    STD. WT.       ERROR

        2      6.45846444     -0.00081930
        3    -14.44931757     -0.00036750
        4      9.10771251     -0.00000021


VAR. NUMBER       WEIGHT

        2          4.13474667
        3         -0.65079751
        4          0.03369741

CONSTANT=         -2.27965415


.....MODEL 2      0.000010
 CRITERION     1
PREDICTORS      2 -   3


   RSQ = 0.76336128                                          3


VAR. NUMBER    STD. WT.       ERROR

        2     -0.41411141      0.00000000
        3      1.27534974      0.00000000
```

```
VAR. NUMBER       WEIGHT

       2        -0.26511654
       3         0.05744177

CONSTANT=        5.35613072


.....MODEL 3      0.000010
 CRITERION    1
PREDICTORS     2 -   2


   RSQ = 0.69742938                                      2


VAR. NUMBER   STD. WT.      ERROR

        2     1.67024475     0.83512238



VAR. NUMBER       WEIGHT

       2         1.06930044

CONSTANT=       -0.66612643

****F-RATIO =  128.8862    D.F. NUM. =   1. D.F. DEN. =  56.      1 0.9283   2 0.7634   PROB = 0.00000

****F-RATIO =   15.8812    D.F. NUM. =   1. D.F. DEN. =  57.      2 0.7634   3 0.6974   PROB = 0.00019

TIME FOR THIS RUN IS      10.62 SECONDS

END-OF-DATA ENCOUNTERED ON SYSTEM INPUT FILE.
```

PROBLEM 6-4

CORRELATIONS

	1	2	3	4
1	1.0000	0.9506	0.9202	0.8608
2	0.9506	1.0000	0.9795	0.9374
3	0.9202	0.9795	1.0000	0.9877
4	0.8608	0.9374	0.9877	1.0000

..... MODEL 1 0.000010
CRITERION 1
PREDICTORS 2 - 4

RSQ = 0.97046899 34

VAR. NUMBER	STD. WT.	ERROR
2	-2.86852345	0.00000089
3	9.18074620	-0.00061560
4	-5.51925820	-0.00102473

VAR. NUMBER	WEIGHT
2	-3.20194390
3	0.71592417
4	-0.03521136
CONSTANT=	5.15428925

..... MODEL 2 0.000010
CRITERION 1
PREDICTORS 2 - 3

RSQ = 0.90655614 3

VAR. NUMBER	STD. WT.	ERROR
2	1.21349736	-0.00000001
3	-0.26840134	-0.00000003

```
VAR. NUMBER       WEIGHT

       2        1.35454722
       3       -0.02093022

CONSTANT=      -2.80357862


 .....  MODEL 3    0.000010
  CRITERION     1
 PREDICTORS      2 -   2


   RSQ = 0.90363374                                        2


 VAR. NUMBER   STD. WT.      ERROR

        2     1.90119304      0.95059652



 VAR. NUMBER       WEIGHT

       2         2.12217665

CONSTANT=       -9.14543152

 ****F-RATIO =  121.1987    D.F. NUM. =    1. D.F. DEN. =  56.      1 0.9705   2 0.9066   PROB = 0.00000

 ****F-RATIO =    1.7826    D.F. NUM. =    1. D.F. DEN. =  57.      2 0.9066   3 0.9036   PROB = 0.13713

TIME FOR THIS RUN IS      10.63 SECONDS

END-OF-DATA ENCOUNTERED ON SYSTEM INPUT FILE.
```

PROBLEM 6-5

CORRELATIONS

	1	2	3	4	5	6	7	8	9
1	1.0000	-0.4941	0.4941	-0.0064	-0.3701	0.3802	-0.3761	0.3811	-0.0234
2	-0.4941	1.0000	-1.0000	0.0519	0.8276	-0.8194	0.6537	-0.6475	0.0540
3	0.4941	-1.0000	1.0000	-0.0519	-0.8276	0.8194	-0.6537	0.6475	-0.0540
4	-0.0064	0.0519	-0.0519	1.0000	0.4432	0.3583	0.5586	0.4857	0.9732
5	-0.3701	0.8276	-0.8276	0.4432	1.0000	-0.6781	0.9540	-0.5359	0.4438
6	0.3802	-0.8194	0.8194	0.3583	-0.6781	1.0000	-0.5357	0.9564	0.3356
7	-0.3761	0.6537	-0.6537	0.5586	0.9540	-0.5357	1.0000	-0.4233	0.5887
8	0.3811	-0.6475	0.6475	0.4857	-0.5359	0.9564	-0.4233	1.0000	0.4831
9	-0.0234	0.0540	-0.0540	0.9732	0.4438	0.3356	0.5887	0.4831	1.0000

```
.....MODEL  1     0.000010
 CRITERION    1
PREDICTORS     2 -   3
               5 -   8

  RSQ = 0.89339033                          425
```

VAR. NUMBER	STD. WT.	ERROR
2	-3.41371149	-0.00006018
3	0.00000000	0.00006018
5	4.30865639	-0.00176566
6	-4.40041941	0.00000204
7	-3.19739789	0.00151154
8	3.33827242	0.00366842

VAR. NUMBER	WEIGHT
2	-15.66179061
3	0.00000000
5	3.02961633
6	-3.22251531
7	-0.26737432
8	0.30236280

CONSTANT= 13.02488339

```
.....MODEL  2     0.000010
 CRITERION    1
PREDICTORS     2 -   3
               5 -   6
```

```
RSQ = 0.25069284                                    5


VAR. NUMBER   STD. WT.      ERROR

      2     -0.65779916     0.00007512
      3      0.00000000    -0.00007512
      5      0.12405860     0.00039836
      6     -0.07462938     0.00000000


VAR. NUMBER      WEIGHT

      2        -3.01792133
      3         0.00000000
      5         0.08723136
      6        -0.05465259

CONSTANT=       6.68037760


.....MODEL  3    0.000010
 CRITERION    1
PREDICTORS     2 -   4


RSQ = 0.24445891                                    3


VAR. NUMBER   STD. WT.      ERROR

      2     -0.49505391     0.00000002
      3      0.00000000    -0.00000002
      4      0.01928963     0.00000020


VAR. NUMBER      WEIGHT

      2        -2.27126127
      3         0.00000000
      4         0.01722990

CONSTANT=       6.31155306


.....MODEL  4    0.000010
 CRITERION    1
PREDICTORS     4 -   4
```

```
   RSQ = 0.00004112                                          2


 VAR. NUMBER   STD. WT.      ERROR

         4    -0.01282496    -0.00641248



 VAR. NUMBER       WEIGHT

        4        -0.01145553

CONSTANT=         5.32699907


 .....MODEL  5     0.000010
  CRITERION     1
 PREDICTORS      2 -    4
                 9 -    9

   RSQ = 0.24894586                                          7


 VAR. NUMBER   STD. WT.      ERROR

         2    -0.49405587    -0.00002122
         3     0.00000000     0.00002122
         4     0.30199035    -0.00003155
         9    -0.29057900    -0.00000000



 VAR. NUMBER       WEIGHT

        2        -2.26668233
        3         0.00000000
        4         0.26974412
        9        -0.02348265

CONSTANT=         5.78559321


 .....MODEL  6     0.000010
  CRITERION     1
 PREDICTORS      4 -    4
                 9 -    9
```

```
RSQ = 0.00558932                                   3


VAR. NUMBER   STD. WT.      ERROR

       4      0.30866408    -C.0C000001
       9     -0.32376140    -C.00000000


VAR. NUMBER        WEIGHT

       4        0.27570523
       9       -0.02616423

CONSTANT=       4.71292442

****F-RATIO =   90.5041   D.F. NUM. =  5. D.F. DEN. =  54.    1 0.8934  90 0.0000  PROB = 0.00000

****F-RATIO =  162.7698   D.F. NUM. =  2. D.F. DEN. =  54.    1 0.8934   2 0.2507  PROB = 0.00000

****F-RATIO =    0.4659   D.F. NUM. =  1. D.F. DEN. =  56.    2 0.2507   3 0.2445  PROB = 0.49769

****F-RATIO =   18.4395   D.F. NUM. =  1. D.F. DEN. =  57.    3 0.2445   4 0.0000  PROB = 0.00007

****F-RATIO =  163.2122   D.F. NUM. =  2. D.F. DEN. =  54.    1 0.8934   5 0.2489  PROB = 0.00000

****F-RATIO =   18.1451   D.F. NUM. =  1. D.F. DEN. =  56.    5 0.2489   6 0.0056  PROB = 0.00008

TIME FOR THIS RUN IS      13.57 SECONDS

END-OF-DATA ENCOUNTERED ON SYSTEM INPUT FILE.
```

PROBLEM 8-1

CORRELATIONS

	1	2	3	4
1	1.0000	-0.0435	1.0000	-0.0541
2	-0.0435	1.0000	-0.0435	0.9800
3	1.0000	-0.0435	1.0000	-0.0541
4	-0.0541	0.9800	-0.0541	1.0000

```
.....MODEL  1     0.000010
 CRITERION    2
PREDICTORS     1 -   1
               4 -   4

   RSQ = 0.96040391                                  3
```

VAR. NUMBER	STD. WT.	ERROR
1	0.00951862	0.00000080
4	0.98047106	0.00000001

VAR. NUMBER	WEIGHT
1	0.00025125
4	10.26054811
CONSTANT=	10.17865217

```
.....MODEL  2     0.000010
 CRITERION    2
PREDICTORS     4 -   4

   RSQ = 0.96031357                                  2
```

VAR. NUMBER	STD. WT.	ERROR
4	1.95991181	0.97995590

```
VAR. NUMBER       WEIGHT

       4       20.51031423
CONSTANT=        8.82680929

****F-RATIO =     0.0890    D.F. NUM. =   1. D.F. DEN. =  39.      1 0.9604   2 0.9603   PROB = 0.75706

****F-RATIO =  967.9013    D.F. NUM. =   1. D.F. DEN. =  40.      2 0.9603  90 0.0000   PROB = 0.00000

TIME FOR THIS RUN IS       8.02 SECONDS

END-OF-DATA ENCOUNTERED ON SYSTEM INPUT FILE.
```

Appendix B

Program Linear IBM 7040 (Fortran IV)

Program Linear IBM 360 (Fortran H)

The material contained in Appendix B is intended to facilitate the computerization of the multiple linear approach.

There are two basic computer programs available for implementation. The first program contains the version utilized in the present text, which is adaptable on IBM 7040, CDC 1604, and like installations. The second program contains the version now operational on the larger IBM 360 series. There are minor differences between these two versions, but the essence of the program is the same. Some computer installations may have other similar multiple regression programs already available. We have tried to make the discussion in this text general enough so that these differences will be minimal.

Recently a special interest group has been formed in the American Educational Research Association in order to coordinate discussion among persons intrigued with the technique. Individuals in various disciplines on many university campuses across the U. S. and abroad are beginning to take advantage of the flexibility which this technique offers.

There are a few differences between program LINEAR and the program MULR04 which follows it. These differences are as follows:

1. End-of-data cards. Since the number of subjects is not specified on the parameter card, a blank card must be inserted after the last subject card. If there are <u>two</u> cards per subject (observation), then there should be <u>two</u> blank cards, etc.

2. Model Card. Program demands two model cards for each model. The second card is simply a continuation of the first card. If a second card is not actually needed, a blank card should be inserted. (Information on the second model card starts in column 22.)

3. Parameter Card:

columns 1-5 = the number of format cards

6-10 = number of variables formated (read off data cards)

11-15 = total number of variables

16-20 = if non-zero, the means, standard deviations, and correlation matrix will be punched.

4. F-Ratio Card in columns 11-13: If this number is "8," then columns 22-25 and columns 26-29 are read in as the number of linearly independent vectors in the full and restricted model, respectively. If columns 11-13 is some other even number (must be greater than 5), then columns 22-25 and columns 26-29 are read as df_1 and df_2, as in program LINEAR.

DATRAN is called DFTRAN in program MULR04.

DFTRAN:

1. Includes the check for the blank card(s) following the last data card.

2. Includes an option to check a subject for missing data (if there is missing data, the subject is skipped).

Data transformation statements are slightly different in DFTRAN. For example, if one wished to generate a vector 10, a product of vectors 2 and 4, the DATRAN statement is: A(J+10)=A(I+2)*A(I+4). For DFTRAN the statement is: X(10)=X(2)*X(4).

```
$IBSYS
$JOB            7044      70 241186 50210 3 MCNEIL
$TIME          10
$IBJOB         NODECK
$IBFTC ALPHA
C      ALPHA               27 SEPTEMBER 1963
C      GENERATES ALPHA INFORMATION
       SUBROUTINE ALPHA(ALPH,NUMHOL,HOLLER)
       DIMENSION FMT1(22),KFMT1(22),FMT2(22),KFMT2(22),A(1),KA(1)
       COMMON FMT1,FMT2,A
       EQUIVALENCE (FMT1,KFMT1),(FMT2,KFMT2),(A,KA)
       DIMENSION ALPH(1),HOLLER(1)
   10 M=(NUMHOL +5)/6
   20 DO 30 I=1,M
   30 ALPH(I)=HOLLER(I)
   40 RETURN
   50 END
$IBFTC BASHLN
C      BASHLN               8 JULY 1964
       SUBROUTINE BASHLN(X,B)
       DIMENSION FMT1(22),KFMT1(22),FMT2(22),KFMT2(22),A(1),KA(1)
       COMMON FMT1,FMT2,A
       EQUIVALENCE (FMT1,KFMT1),(FMT2,KFMT2),(A,KA)
       B = 0.
       IF(X-2.)3,1,1
     1 Z = 1./(X*X)
     2 B = (X+.5)*ALOG(X)-X+.918938534+(.833333333E-1-Z
      1*(.277777778E-2-Z*(.793650794E-3-Z*(.595238095E-
      23-Z*.841750842E-3))))/X
     3 RETURN
     4 END
$IBFTC BTALOG
C      BTALOG               8 JULY 1964
       SUBROUTINE BTALOG(AG,BG,CG,X)
       DIMENSION FMT1(22),KFMT1(22),FMT2(22),KFMT2(22),A(1),KA(1)
       COMMON FMT1,FMT2,A
       EQUIVALENCE (FMT1,KFMT1),(FMT2,KFMT2),(A,KA)
     1 CALL GAMLOG(AG,S)
     2 CALL GAMLOG(BG,T)
     3 CALL GAMLOG(CG,U)
     4 X = S+T-U
     5 RETURN
     6 END
$IBFTC CORRLB
C      CORRLB               8 FEBRUARY 1965
C      MEANS,STANDARD DEVIATIONS,CORRELATIONS
C      USES SIGMA AS BUFFER FROM TAPE
       SUBROUTINE CORRLB(IDTAPE,FILE,NUM,NVAR,LMEAN,LSIGMA,LCORR)
       DIMENSION FMT1(22),KFMT1(22),FMT2(22),KFMT2(22),A(1),KA(1)
       COMMON FMT1,FMT2,A
       EQUIVALENCE (FMT1,KFMT1),(FMT2,KFMT2),(A,KA)
   20 CALL ZEROST(LCORR,NVAR,NVAR)                                     104
   21 CALL ZEROST(LMEAN,NVAR,1)                                        105
   22 CALL ZEROST(LSIGMA,NVAR,1)                                       106
```

```
   23 CALL POSTAP(LOCREC,IDTAPE,0,FILE)
   30 DO  74  I=1,NUM                                                   108
   40 ITO=LSIGMA+NVAR-1                                                 109
   41 READ(IDTAPE)LOCT,(A(II),II=LSIGMA,ITO)
   51 DO  74  J=1,NVAR                                                  112
      IA=LSIGMA-1+J                                                     113
   60 IT=LMEAN-1+J                                                      114
   61 A(IT)=A(IT)+A(IA)                                                 115
   70 DO  74  L=J,NVAR                                                  116
      IB=LSIGMA-1+L                                                     117
   71 IT= (LCORR-1)+((L-1)*NVAR)+J                                      118
   74 A(IT)=A(IT)+A(IA)*A(IB)
   75 REWIND IDTAPE                                                     122
C     COMPUTE R MATRIX                                                  123
   78 CALL ZEROST(LSIGMA,NVAR,1)                                        124
   80 FN=NUM                                                            125
C     COMPLTE NON DIAGONAL ELEMENTS OF R MATRIX                         126
      KM1=NVAR-1                                                        127
   81 DO 130 I=1,KM1                                                    128
   82 IP1=I+1                                                           129
   83 DO 130 J=IP1,NVAR                                                 130
   84 ISI=LCORR-1+(I-1)*NVAR+I                                          131
   85 ISJ=LCORR-1+(J-1)*NVAR+J                                          132
   86 ISIJ=LCORR-1+(J-1)*NVAR+I                                         133
   87 ISJI=LCORR-1+(I-1)*NVAR+J                                         134
   88 IMI=LMEAN-1+I                                                     135
   89 IMJ=LMEAN-1+J                                                     136
   90 DEN=SQRT((FN*A(ISI)-A(IMI)*A(IMI))*(FN*A(ISJ)-A(IMJ)*A(IMJ)))
   95 IF(DEN) 110,100,110                                               138
  100 A(ISIJ)=0.0                                                       139
  105 GO TO 130                                                         140
  110 A(ISIJ)=(FN*A(ISIJ)-A(IMI)*A(IMJ))/DEN                            141
  115 IF(ABS(A(ISIJ))-1.0) 130,130,120
  120 WRITE (6,125) I,J,A(ISIJ)
  125 FORMAT(28H OUT OF RANGE CORRELATION I=I5,5H   J=I5,5H   R=F10.4)  144
  126 CALL EXIT                                                         145
  130 A(ISJI)=A(ISIJ)                                                   146
C     COMPUTE MEAN AND SIGMA                                            147
  135 DO 165 I=1,NVAR                                                   148
  140 IM=LMEAN-1+I                                                      149
  145 A(IM)=A(IM)/FN                                                    150
  150 IS=LSIGMA-1+I                                                     151
  155 II=LCORR-1+(I-1)*NVAR+I                                           152
  160 A(IS)=SQRT((A(II)/FN)-A(IM)*A(IM))
C     COMPUTE DIAGONAL ELEMENTS OF R MATRIX                             154
  161 IF(A(IS))162,162,164                                              155
  162 A(II)=0.0                                                         156
  163 GO TO 165                                                         157
  164 A(II)=1.0                                                         158
  165 CONTINUE                                                          159
  157 CALL PRIMSC(LMEAN,LSIGMA,LCORR,NVAR)
  170 RETURN                                                            160
      END                                                               164
$IBFTC DATMOV
C     DATMOV          8 FEBRUARY 1965
C     MOVES DATA FROM ANY FORTRAN NUMBERED DEVICE 0-8 TO ANOTHER DEVICE
      SUBROUTINE DATMOV(IDFRDV,FRFILE,IFRFMT,NR,NFRCOL,LFRBUF,IDTODV,
     1NUMHOL,TOFILE,ITOFMT,NTOCOL,LTOBUF)
      DIMENSION FMT1(22),KFMT1(22),  FMT2(22),KFMT2(22),A(1),KA(1)
```

```
      COMMON FMT1,FMT2,A
      EQUIVALENCE (FMT1,KFMT1),(FMT2,KFMT2),(A,KA)
      DIMENSION FRFILE(2),TOFILE(22)
      ITYPO=ITOFMT/10
    1 IF(FRFILE(1)) 4 , 2 ,4
    2 INPER=2
    3 GO TO 5
    4 INPER=1
    5 IF(NUMHOL) 8,6,8
    6 IOPER=2
    7 GO TO 9
    8 IOPER=1
    9 GO TO( 20 , 30 ),INPER
   20 CALL POSTAP(KOUNT,IDFRDV,0,FRFILE)
   30 GO TO(40,80),IOPER
   40 IF(NUMHOL-12) 41,45,45
 41   WRITE(6,42) NUMHOL
   42 FORMAT(42H NUMHOL IN DATMOV LESS THAN 12.  NUMHOL = ,I3)
   43 CALL EXIT
   45 CALL POSTAP(KOUNT,IDTODV,999999999)
   50 IRECRD=-KOUNT
   60 NWORDS=(NUMHOL+5)/6
   70 WRITE(IDTODV)IRECRD,ITYPO,NR,NTOCOL,NWORDS,(TOFILE(I),I=1,NWORDS)
   80 ITYPI= IFRFMT/10
   90 INFMT= MOD(IFRFMT,10)
  100 IOFMT= MOD(ITOFMT,10)
C          SET INPUT FORMAT (FMT1) - IF REQUIRED.
  110 IF(ITYPI)200,120,200
  120 IF(INFMT)130,190,130
  130 GO TO(  200, 140 ),INFMT
 140  READ (5,150) (FMT1(I),I=1,14)
  150 FORMAT(13A6,1X,F1.0)
  160 IF(FMT1(14)) 170,200,170
 170  READ(5,150) (FMT1(I),I=14,22)
  180 GO TO 200
  190 GO TO (191,193),INPER
  191 CALL ALPHA(FMT1(1),12,12H(9X,13A6,A2))
  192 GO TO 200
  193 CALL ALPHA(FMT1(1), 9, 9H(13A6,A2))
C          SET OUTPUT FORMAT (FMT2) - IF REQUIRED.
  200 IF(ITYPO)300,220,300
  220 IF(IOFMT)230,290,230
  230 GO TO( 300, 240 ),IOFMT
 240  READ(5,150) (FMT2(I),I=1,14)
  260 IF(FMT2(14))270,300,270
 270  READ(5,150) (FMT2(I),I=14,22)
  280 GO TO 300
  290 GO TO (291,293),IOPER
  291 CALL ALPHA(FMT2(1),12,12H(I9,13A6,A2))
  292 GO TO 300
  293 CALL ALPHA(FMT2(1), 9, 9H(13A6,A2))
C          MISC INITIALIZATION BEFORE MOVE CF DATA STARTED.
  300 KTO2=LTOBUF-1+NTOCOL
  348 KTO2=LTOBUF-1+NTOCOL
```

```
  352 KTO=LFRBUF-1+NFRCOL
C          PROCESS NR RECORDS
  349 DO  560  I=1,NR
C          DATA IN
  410 IF(ITYPI)440,420,440
  420 READ(IDFRDV,FMT1)(A(J),J=LFRBUF,KTO)
  425 GO TO 486
  440 GO TO(450,470),INPER
  450 READ(IDFRDV)K,(A(J),J=LFRBUF,KTO)
  450 GO TO 486
  470 READ(IDFRDV)(A(J),J=LFRBUF,KTO)
C          DATA TRANSFORMATION
  486 CALL DATRAN(LFRBUF-1,NFRCOL,LTOBUF-1,NTOCOL)
C          DATA OUT
  511 KOUNT=KOUNT+1
  512 IF(ITYPO )520,531,520
  520 GO TO(522 ,525),IOPER
  522 WRITE(IDTODV)KOUNT,(A(J),J=LTOBUF,KTO2)
  524 GO TO 560
  525 WRITE(IDTODV)(A(J),J=LTOBUF,KTO2)
  535 GO TO 560
  531 GO TO(532, 534),IOPER
  532 WRITE(IDTODV,FMT2)KOUNT,(A(J),J=LTOBUF,KTO2)
  533 GO TO 560
  534 WRITE(IDTODV,FMT2)(A(J),J=LTOBUF,KTO2)
  560 CONTINUE
C          END OF LOOP TO MOVE DATA
 6001 GO TO(6002,601),IOPER
C          WRITE END LABEL IF IDTODV IS PERSUE FILE.
 6002 KOUNT=KOUNT+1
 6003 I=999999999
 6004 WRITE(IDTODV)I,KOUNT
  601 RETURN
      END
$IBFTC GAMLOG
C     GAMLOG                    8 JULY 1964
      SUBROUTINE GAMLOG(X,AG)
      DIMENSION FMT1(22),KFMT1(22),FMT2(22),KFMT2(22),A(1),KA(1)
      COMMON FMT1,FMT2,A
      EQUIVALENCE (FMT1,KFMT1),(FMT2,KFMT2),(A,KA)
      IF(X.EQ.1.) GO TO 15
    1 J = AMOD(X,2.)+1.
    2 GO TO(7,3),J
    3 T = X-2.
    4 U = (X-3.)*.5
    5 AG = .57236493-T*.69314718I
    6 GO TO 11
    7 T = (X-2.)*.5
    8 AG = 0.
    9 B = 0.
   10 GO TO 12
   11 CALL BASHLN(U,B)
   12 CALL BASHLN(T,C)
   13 AG = AG+C-B
   14 GO TO 16
   15 AG = .57236493
   16 RETURN
   17 END
$IBFTC MOVCOR
```

```
C       MOVCOR                  15 APRIL 1964
C       MOVE DATA IN CORE                                                    321
     10 SUBROUTINE MOVCOR(LFRM,NR,NC,LTO)                                    322
        DIMENSION FMT1(22),KFMT1(22),FMT2(22),KFMT2(22),A(1),KA(1)
        COMMON FMT1,FMT2,A
        EQUIVALENCE (FMT1,KFMT1),(FMT2,KFMT2),(A,KA)
     11 NWORDS=NR*NC
     12 CALL TMT(NWORDS,A(LTO),A(LFRM))
     18 RETURN                                                               333
     19 END                                                                  334
$IBFTC PLEVEL
C       PLEVEL                   8 JULY 1964
        SUBROUTINE PLEVEL(DF1,DF2,F,P)
        DIMENSION FMT1(22),KFMT1(22),FMT2(22),KFMT2(22),A(1),KA(1)
        COMMON FMT1,FMT2,A
        EQUIVALENCE (FMT1,KFMT1),(FMT2,KFMT2),(A,KA)
        AX = DF2/((F*DF1)+DF2)
         C = DF1+DF2
        CALL BTALOG(DF1,DF2,C,BETA)
        CALL PROBLY(AX,BETA,DF1,DF2,P)
        RETURN
        END
$IBFTC POSTAP
C       POSTAP                  8 FEBRUARY 1965
C       POSITIONS TAPE READY TO READ OR WRITE RECORD
        SUBROUTINE POSTAP (LOCREC,IDTAPE,IDREC,KFILE)
        DIMENSION FMT1(22),KFMT1(22),FMT2(22),KFMT2(22),A(1),KA(1)
        COMMON FMT1,FMT2,A
        EQUIVALENCE (FMT1,KFMT1),(FMT2,KFMT2),(A,KA)
        DIMENSION KFILE(2)
     10 REWIND IDTAPE
     11 IRECRD=1
     12 FORMAT(I9,13A6,A2)
     13 LAST=0
     20 IF(IRECRD) 21, 23,21
     21 READ (IDTAPE)I
     22 GO TO 30
     23 READ (IDTAPE,12)I
     24 IF(I-LAST)30,25,30
     25 IRECRD=1
     30 IF(I)50 ,110,90
     50 BACKSPACE IDTAPE
     60 READ(IDTAPE) I,ITYPE,NR,NC,NWORDS,KNAME1,KNAME2
     70 IF(IDREC) 110, 71,104
     71 IF(KFILE(1)-KNAME1) 75, 72, 75
     72 IF(KFILE(2)-KNAME2) 75, 73, 75
     73 LOCREC=-I+1
     74 GO TO 106
     75 IF(ITYPE) 20, 76,20
     76 LAST=-I+NR
     77 IRECRD=0
     78 GO TO 20
     90 IF (I-999999999)103,100,110
    100 IF (IDREC-999999999)110,101,110
```

```
 101  BACKSPACE IDTAPE
 102  READ(IDTAPE) I,J
     I=J
      GO TO 105
 103 IF(IDREC-I) 20,105,20
 104 IF(IDREC-IABS(I)) 75,105, 75
 105 LOCREC=I
     BACKSPACE IDTAPE
 106 RETURN
 110  WRITE (6,111) IDTAPE,IDREC,KFILE(1),KFILE(2)
 111 FORMAT(35H ERROR IN POSTAP, ARGUMENTS-IDTAPE=I2,7H IDREC= I5,6H FIL
    1E=2X2A6)
 112 CALL EXIT
     END
$IBFTC PRIMSC
C      PRIMSC              29 SEPTEMBER 1963
C      PRINTS MEANS,STANDARD DEVIATIONS, CORRELATIONS
       SUBROUTINE PRIMSC(LMEAN,LSIGMA,LCORR,NVAR)
       DIMENSION FMT1(22),KFMT1(22),FMT2(22),KFMT2(22),A(1),KA(1)
       COMMON FMT1,FMT2,A
       EQUIVALENCE (FMT1,KFMT1),(FMT2,KFMT2),(A,KA)
  20   WRITE (6,30)
   30  FORMAT (41H1  MEANS-STANDARD DEVIATICNS-CORRELATIONS )
  40   WRITE (6,50)
   50  FORMAT(/////10H     MEANS)                                         410
   60  CALL PRINT (LMEAN,1,NVAR,1,0)
  70   WRITE (6,80)
   80  FORMAT(/////24H      STANDARD DEVIATIONS)                          413
   90  CALL PRINT (LSIGMA,1,NVAR,1,0)
 100   WRITE (6,110)
  110  FORMAT(17H1     CORRELATIONS)
  120  CALL PRINT (LCORR,NVAR,NVAR,1,0)
  121  RETURN                                                             418
  122  END                                                                419
$IBFTC PRINT
C      PRINT                15 APRIL 1964
C      PRINTS MATRIX PARTIONED BY COLUMNS USING FORMAT IN FMT2
       SUBROUTINE PRINT(LARRAY,NR,NC,IDFMT,NUMHOL,HOLLER)
       DIMENSION FMT1(22),KFMT1(22),FMT2(22),KFMT2(22),A(1),KA(1)
       COMMON FMT1,FMT2,A
       EQUIVALENCE (FMT1,KFMT1),(FMT2,KFMT2),(A,KA)
       DIMENSION HOLLER(22)
    1  IF(NUMHOL) 30, 30,10
   10  N=(NUMHOL+5)/6
  20   WRITE (6,21) (HOLLER(I),I=1,N)
   21  FORMAT (////1X,21A6,A5///)
   30  IF (IDFMT)35,60,35
   35  GO TO (40,51,55,59),IDFMT
   40  CALL ALPHA (FMT2,15,15H(1H I9,10F12.4))
   50  GO TO 74
   51  CALL ALPHA (FMT2,13,13H(1H I9,10I12))
   52  GO TO 74
   55  CALL ALPHA (FMT2,12,12H(1H I9,20A6))
   56  KSIZE=20
   57  CALL ALPHA(FMT1,19,19H(///8HS       20I6))
   58  GO TO 76
   59  CALL ALPHA (FMT1,18,18H(///6HS     ,6I20))
  500  CALL ALPHA (FMT2,14,14H(1H I9,6E20.8))
  501  KSIZE=6
```

```
 502 GO TO 76
 60  READ (5,61) (FMT2(I),I=1,15)
  61 FORMAT (13A6,A1,F1.0)
  62 IF (FMT2(15)) 63,74,63
 63  READ (5,61) (FMT2(I),I=15,22)
  74 KSIZE=10
  75 CALL ALPHA(FMT1,20,20H(///8HS        10I12))
  76 KMANY=(((NC -1)/KSIZE)+1)*KSIZE
  80 DO 210  L=KSIZE,KMANY,KSIZE
  90 L1=L-(KSIZE-1)
 100 IF(L-KMANY)130,110,130
 110 L2=NC
 120 GO TO 140
 130 L2=L
140  WRITE (6,FMT1) (I,I=L1,L2)
 150 DO 210 I=1,NR
 160 KSTAR=LARRAY+L-KSIZE+(I-1)*NC
 170 IF(L-KMANY)180,200,180
 180 KSTOP=KSTAR+(KSIZE-1)
 190 GO TO 210
 200 KSTOP=LARRAY-1+I*NC
 210 WRITE (6,FMT2) I,(A(J),J=KSTAR,KSTOP)
 220 RETURN
 230 END
$IBFTC PROBLY
C      PROBLY                  8 JULY 1964
       SUBROUTINE PROBLY(AX,B,DF1,DF2,P)
       DIMENSION FMT1(22),KFMT1(22),FMT2(22),KFMT2(22),A(1),KA(1)
       COMMON FMT1,FMT2,A
       EQUIVALENCE (FMT1,KFMT1),(FMT2,KFMT2),(A,KA)
       IF(DF1.LE.0..OR.DF2.LE.0..OR.AX.GE.1.)GO TO 11
       DIMENSION Y(6)
       Y(1) =   .4054651181
       Y(2) = -1.203972814
       Y(3) =   .5877866649
       Y(4) =  Y(2)
       Y(5) =  Y(1)
       Y(6) = -.5108256238
       C = B-69.-.693147181
       CM = (DF1-1.)*.5
       CN = (DF2-1.)
        H = ARSIN(SQRT(AX))/60.
        P = 0.
    5   X = 0.
       DO 9 I=1,10
       GO TO(7,7,7,7,7,7,7,7,7,6),I
    6 Y(6)= Y(2)
    7 DO 9 J=1,6
      X = X+H
      XS = SIN(X)
       Z = Y(J)-C+CN*ALOG(XS)+CM*ALOG(1.-XS*XS)
      IF(Z)9,8,8
    8  P = P+EXP(Z-69.)
    9 CONTINUE
```

```
      P = P*H
   10 RETURN
   11 WRITE (6,12) DF1,DF2,AX
   12 FORMAT(1H1,6HDF1 = F5.0,7H,DF2 = F5.0,6H,AX = F6.0//4
     18H P DOES NOT EXIST IF DF1 OR DF2 IS LESS THAN 1  /4
     28H OR IF AX IS GREATER THAN 1(WHICH IMPLIES THAT F/4
     38H IS NEGATIVE). CONTROL IS THEREFORE RETURNED TO /4
     48H THE MONITOR.                                    )
      CALL EXIT
      END
$IBFTC TAPGEN
C     TAPGEN               26 SEPTEMBER 1963
C     GENERATES A TAPE READY FOR USE BY OTHER TAPE ROUTINES
      SUBROUTINE TAPGEN (IDTAPE)
      DIMENSION FMT1(22),KFMT1(22),FMT2(22),KFMT2(22),A(1),KA(1)
      COMMON FMT1,FMT2,A
      EQUIVALENCE (FMT1,KFMT1),(FMT2,KFMT2),(A,KA)
   10 I=999999999
   11 J=1
   20 REWIND IDTAPE
   30 WRITE (IDTAPE) I,J
   40 REWIND IDTAPE
      RETURN
      END
$IBFTC REGRED
C     REGRED               9 FEBRUARY 1965
C     ITERATIVE REGRESSION
   10 SUBROUTINE REGRED(LMEAN,LSIGMA,LCORR,LSTDWT,LWTS,LRSQ,NVAR)          47
      DIMENSION FMT1(22),KFMT1(22),FMT2(22),KFMT2(22),A(1),KA(1)
      COMMON FMT1,FMT2,A
      EQUIVALENCE (FMT1,KFMT1),(FMT2,KFMT2),(A,KA)
      DIMENSION MFLD(55),MFLDL(27),PROBID(2)
      K6=0
 31   READ (5,32) PROBID,NFLDS,STOPC,(MFLD(I),I=1,NFLDS)
   32 FORMAT(2A5,I3,E8.1,28I2/21X,28I2)
      IF(NFLDS) 35,35,36
   35 RETURN
   36 IF(NFLDS-6) 37,400,37
   37 K5=NFLDS-1
   38 IDC=MFLD(NFLDS)
   40 WRITE (6,41) PROBID,STOPC,IDC,(MFLD(I),I=1,K5)
   41 FORMAT(///2X,5H.....2A5,F10.6/12H    CRITERIONI5/12H  PREDICTORSI6,
     12H -I4/(12X,I6,2H -I4))
C
  120 NFLD1=NFLDS-1                                                       490
  130 DO    160 I=2,NFLD1,2                                               491
  140 M=I/2                                                               492
  150 MFLDL(M)=MFLD(I)                                                    493
  160 MFLD(M)=MFLD(I-1)                                                   494
C     INITIALIZE                                                          495
  170 CALL ZEROST(LWTS,NVAR,1)                                            496
  180 CALL ZEROST(LSTDWT,NVAR,1)                                          497
  190 S=0.0                                                               498
  200 SIG2=0.0                                                            499
  201 RSQ=0.0                                                             500
  202 DEL=0.0                                                             501
  210 ITER=0                                                              502
  211 ID=1                                                                503
  212 NGRP=NFLDS/2                                                        504
```

```
C 213 SET FOR NEW ITERATION                                  505
  220 RSQL=0.0                                               506
  230 DO 255 I=1,NGRP                                        508
C 221 ITERATE                                                507
  231 KSTAR=MFLD(I)                                          509
  232 KSTOP=MFLDL(I)                                         510
  233 DO  255  J=KSTAR,KSTOP                                 511
  234 IA=(LWTS-1)+J                                          512
  235 IB=(LCORR-1)+((ID-1)*NVAR)+J                           513
  236 IC=(LCORR-1)+((IDC-1)*NVAR+J)                          514
  237 A(IA)=A(IA)+(DEL*A(IB))                                515
  238 DEN=S-(A(IA)*A(IC))                                    516
  239 IF(DEN)245,240,245                                     517
  240 DELT= A(IC)                                            518
  241 STEST=DELT*DELT                                        519
  242 SIG2T=STEST                                            520
  243 RSQT=STEST                                             521
  244 GO TO 249                                              522
  245 DELT=((SIG2*A(IC))-(S*A(IA)))/DEN                      523
  246 STEST=S+(DELT*A(IC))                                   524
  247 SIG2T=SIG2+(2.0*A(IA)*DELT)+(DELT*DELT)                525
  248 RSQT=(STEST*STEST)/SIG2T                               526
  249 IF(RSQL-RSQT)250,255,255                               527
  250 SLAR=STEST                                             528
  251 SIG2L=SIG2T                                            529
  252 RSQL=RSQT                                              530
  253 DELTL=DELT                                             531
  254 IDLAR=J                                                532
  255 CONTINUE                                               533
 1255 IF(RSQL-RSQ-STOPC) 268,256,256
  256 S=SLAR                                                 535
  257 SIG2=SIG2L                                             536
  258 RSQ=RSQL                                               537
  259 DEL=DELTL                                              538
  260 ITER=ITER+1                                            539
  261 ID= IDLAR                                              540
  262 IA = (LSTDWT-1) + ID                                   541
  263 A(IA)=A(IA)+DEL                                        542
 1265 IF(RSQ-1.) 220,268,1266
 1266 WRITE(6,1267)
 1267 FORMAT(25H RSQ IS GREATER THAN ONE.)                   547
 1268 CALL EXIT                                              548
C 267 TERMINATE                                              549
  268 IF(ITER) 500,500,1269
 1269 SDS2=S/SIG2
 2268 WRITE(6,265) RSQL,ITER
  265 FORMAT(//5X,5HRSQ = F11.8, 40X, I5)
  269 DO  274    I=1,NGRP                                    551
  270 KSTAR=MFLD(I)                                          552
  271 KSTOP=MFLDL(I)                                         553
  272 DO  274 J= KSTAR,KSTOP                                 554
  273 IA=LSTDWT-1+J                                          555
  274 A(IA)=A(IA)*SDS2                                       556
  275 WRITE (6,276)
```

```
 276 FORMAT(///)                                                      558
 277 WRITE (6,278)
 278 FORMAT(34H  VAR. NUMBER   STD. WT.     ERROR//)                 560
 279 DO 295 I=1,NGRP                                                  561
 280 KSTAR=MFLD(I)                                                    562
 281 KSTOP=MFLDL(I)                                                   563
 282 DO 295  J=KSTAR,KSTOP                                            564
 283 IA=LWTS-1+J                                                      565
 284 A(IA)=0.0                                                        566
 285 DO  291  IL=1,NGRP                                               567
 286 LSTAR=MFLD(IL)                                                   568
 287 LSTOP=MFLDL(IL)                                                  569
 288 DO  291  L=LSTAR,LSTOP                                           570
 289 IB=LSTDWT-1+L                                                    571
 290 IC=LCORR-1+J+((L-1)*NVAR)                                        572
 291 A(IA)=A(IA)+(A(IB)*A(IC))                                         573
 292 IC=LCORR-1+J+((IDC-1)*NVAR)                                      574
 293 A(IA)=A(IA)-A(IC)                                                575
 294 IB=LSTDWT-1+J                                                    576
295  WRITE (6,296) J,A(IB),A(IA)
 296 FORMAT(1H I10,F15.8,F15.8)                                       578
C 297 COMPUTE REGRESSION EQUATION                                     579
 298 WRITE (6,299)
 299 FORMAT(/////25H  VAR. NUMBER      WEIGHT//)                      581
 300 FK1=0.0                                                          582
 301 DO  315 I=1,NGRP                                                 583
 302  KSTAR=MFLD(I)                                                   584
 303 KSTOP=MFLDL(I)                                                   585
 304 DO  315  J=KSTAR,KSTOP                                           586
 305 IA=LSIGMA-1+J                                                    587
 306 IB=LSTDWT-1+J                                                    588
 307 IC=LSIGMA-1+IDC                                                  589
 308 ID=LMEAN-1+J                                                     590
 309 IE=LWTS-1+J                                                      591
 310 IF(A(IA))313,311,313                                             592
 311 A(IE)=0.0                                                        593
 312 GO TO 315                                                        594
 313 A(IE)=A(IB)*(A(IC)/A(IA))                                        595
 314 FK1=FK1+(A(IB)*(A(ID)/A(IA)))                                    596
 315 WRITE (6,316) J,A(IE)
 316 FORMAT(1H I9,F18.8)                                              598
 317 ID = LMEAN-1+ IDC                                                599
 318 REGCO=A(ID)-(A(IC)*FK1)                                          600
 321 WRITE (6,330) REGCO
 330 FORMAT (/10H CONSTANT=F18.8)
 323 LAPN=LWTS+NVAR                                                   605
 324 A(LAPN)=REGCO                                                    606
 325 K5=LRSQ+K6
 326 A(K5)=RSQL
 327 K6=K6+1
 328 GO TO 31
C          COMPUTE F RATIO
 400 DF1=MFLD(3)*100+MFLD(4)
 401 DF2=MFLD(5)*100+MFLD(6)
 402 K8=MFLD(1)-1+LRSQ
 403 K9=MFLD(2)-1+LRSQ
 404 F=((A(K8)-A(K9))/DF1) / ((1.0-A(K8))/DF2)
 410 CALL PLEVEL(DF1,DF2,F,P)
 405 WRITE(6,406) F,DF1,DF2,MFLD(1),A(K8),MFLD(2),A(K9),P
```

```
  406 FORMAT(//2X,13H****F-RATIO =F10.4,3X,12H D.F. NUM. =F5.0,12H D.F.
     1DEN. =F5.0,3X,I4,F7.4,I4,F7.4,3X,6HPROB =,F8.5)
  407 GO TO 31
C
  500 WRITE (6,501)
  501 FORMAT(58H REGRED EXIT.  STOPC GREATER THAN RSQ FOR FIRST ITERATIO
     1N.)
  502 CALL EXIT
  408 END
$IBFTC ZEROST
C      ZERO  PART OF STORAGE                                                  678
   10 SUBROUTINE ZEROST (LOC,NR,NC)                                           679
      DIMENSION FMT1(22),KFMT1(22),FMT2(22),KFMT2(22),A(1),KA(1)
      COMMON FMT1,FMT2,A
      EQUIVALENCE (FMT1,KFMT1),(FMT2,KFMT2),(A,KA)
   11 KEND=(LOC-1)+(NR*NC)                                                    683
   12 DO  13   I=LOC,KEND                                                     684
   13 A(I)=0.0                                                                685
   14 RETURN                                                                  686
   15 END                                                                     687
$IBMAP TMT
*
*      CALL TMT (NR OF WORDS, TO ADDRESS, FROM ADDRESS)
*
LAST    SXA     *+1,4
        TMT     **
XR4     AXT     **,4
TMT     TRA     **
        SXA     XR4,4
        LAC     TMT,4
        CAL     3,4
        LGR     18
        CAL     4,4
        LGR     18
        CAL*    2,4
        PAX     ,4
        LGL     36
        TNX     LAST,4,255
        TMT     255
        TRA     *-2
        ENTRY   TMT
        END
$IBFTC DATRAN  NODECK
      SUBROUTINE DATRAN(I,NFRCOL,J,NTOCOL)
C     DATRAN              7 JANUARY 1964          PAGE 1
C     TRANSFORMS DATA MOVING FROM TPTOTP
      DIMENSION FMT1(22),KFMT1(22),FMT2(22),KFMT2(22),A(1),KA(1)
      COMMON FMT1,FMT2,A
      EQUIVALENCE (FMT1,KFMT1),(FMT2,KFMT2),(A,KA)
      CALL ZEROST (J+1,1,NTOCOL)
   10 CALL MOVCOR(I+1, 1, NFRCOL, J+1)
   20 RETURN
   30 END
$IBFTC LINEAR  NODECK
```

```
C      PROGRAM LINEAR WITH TIMING AND USE OF ONE DISK UNIT
       DIMENSION FMT1(22),FMT2(22),KFMT1(22),KFMT2(22),A(4000),KA(6000)
       COMMON FMT1,FMT2,A,IT(1)
       EQUIVALENCE (FMT1,KFMT1),(FMT2,KFMT2),(A,KA)
     1 CALL TAPGEN(8)
       ITIME = IT(8)
       READ (5,5) (FMT1(I),I=1,12)
       IF (FMT1(1) - FMT1(2)) 10, 15, 10
     5 FORMAT (12A6)
    10 WRITE (6,20) (FMT1(I),I=1,12)
    20 FORMAT (1H1, 5X, 12A6////)
       READ (5,25) NC,NIV,NTV
    25 FORMAT (16I5)
       CALL DATMOV(5,0,2,NC,NIV,1,8,12,12H DATA         ,10,NTV,101)
       CALL CORRLB (8,12H DATA        ,NC,NTV,1,101,201)
       CALL ZEROST(10401,1,100)
       CALL REGRED(1,101,201,10201,10301,10401,NTV)
       TM = IT(8) - ITIME
       TM = TM/60.
       WRITE(6,30) TM
    30 FORMAT(22H-TIME FOR THIS RUN IS ,F10.2,8H SECONDS  )
       GO TO 1
    15 CALL EXIT
       END
$ENTRY          LINEAR
$IBSYS
```

```
LEVEL 13  (23 MAY 67)                        OS/360  FORTRAN H                                    DATE  68.196/19.26.08

       COMPILER OPTIONS - NAME=  MR04,OPT=02,LINECNT=59,SOURCE,EBCDIC,NOLIST,NODECK,LOAD,NOMAP,NOEDIT,ID
C      MULR04          DIVISION OF EDUCATIONAL RESEARCH SERVICES
C                                UNIVERSITY OF ALBERTA
C      ............................................................
C
C      PURPOSE:
C         CALCULATES CORRELATIONS AND CARRIES OUT REGRESSION ANALYSIS.
C         FOR EACH REGRESSION MODEL, THE PROGRAM COMPUTES THE SQUARED
C         MULTIPLE CORRELATION AND THE REGRESSION WEIGHTS BOTH
C         STANDARDIZED AND RAW SCORE FORM FOR EACH VARIABLE SPECIFIED AS
C         A PREDICTOR.
C
C      CARD INPUT:
C         1. TITLE  (20A4)
C         2. PARAMETERS  (4I5)
C         3. FORMAT OF DATA  (20A4)
C         4. DATA  (ACCORDING TO ABOVE FORMAT)
C         5. END OF DATA CARD(S)    AS REQUIRED BY DFTRAN
C                 THERE MUST BE AS MANY END-OF-DATA CARDS AS THERE ARE
C                 DATA CARDS PER OBSERVATION
C         6. MODEL CARDS  (5A2,I3,E8.1,28I2,/,21X28I2)
C         7. F-RATIO CARDS  (5A2,I3,8X4I2)
C         8. BLANK CARD  (INDICATES END OF MODEL AND F-RATIO CARDS)
C
C      DESCRIPTION OF PARAMETERS:
C         NFMT   - NUMBER OF FORMAT CARDS
C                  (MAX. 4, AT LEAST ONE ASSUMED)
C         NVARIN - NUMBER OF VARIABLES INPUT
C         NVART  - TOTAL NUMBER OF VARIABLES AFTER TRANSFORMATION
C         IPUNCH - IF NON-ZERO, MEANS, STANDARD DEVIATIONS AND
C                  CORRELATIONS PUNCHED ON CARDS IN BINARY FORM
C
C      DESCRIPTION OF MODEL AND F-RATIO CARDS:
C         MODEL CARDS - PROBID, NFLDS, STOPC, (MFLD(I),I=1,NFLDS)
C              PROBID - IDENTIFICATION OF MODEL.  MODELS SHOULD BE
C                       NUMBERED CONSECUTIVELY.
C              NFLDS  - NUMBER OF ELEMENTS IN MFLD.  THIS MUST BE AN ODD
C                       NUMBER
C              STOPC  - ITERATION STOP CRITERION.  ITERATION WILL
C                       CONTINUE UNTIL THE SQUARED MULTIPLE CORRELATION
C                       ON TWO SUCCESSIVE ITERATIONS DIFFERS BY LESS THAN
C                       STOPC.
C              MFLD   - VECTOR INDICATING VARIABLES TO BE INCLUDED IN
C                       MODEL.  ALL VARIABLES BETWEEN EACH PAIR OF
C                       ELEMENTS OF MFLD WILL BE INCLUDED IN THE MODEL.
C                       THE LAST ELEMENT IS THE CRITERION.
C        F-RATIO CARDS - PROBID, NFLDS, STOPC, (MFLD(I),I=1,NFLDS)
C              PROBID - IDENTIFICATION OF F-RATIO.
C              NFLDS  - MUST BE AN EVEN NUMBER >5
C              STOPC  - MAY BE LEFT BLANK.
C              MFLD   - MFLD(1) NUMBER CORRESPONDING TO FULL MODEL.
C                       MFLD(2) NUMBER CORRESPONDING TO RESTRICTED MODEL.
C                       MODEL 99 UNLESS OTHERWISE USED HAS A RSQ =0.
C                       MFLD(3) AND MFLD(4) TREATED AS ONE NUMBER
C                       REPRESENTING DF1
C                       MFLD(5) AND MFLD(6) TREATED AS ONE NUMBER
C                       REPRESENTING DF2
C                       IF NFLDS=8, DF1 AND DF2 ARE TAKEN AS THE NUMBER
C                       OF INDEPENDENT WEIGHTS, INCLUDING THE CONSTANT,
```

```
         C                   IN THE FULL AND RESTRICTED MODELS RESPECTIVELY.
         C                   OTHERWISE, THEY ARE RESPECTIVELY THE DEGREES OF
         C                   FREEDOM OF THE NUMERATOR AND DENOMINATOR.
         C
         C     REMARKS:
         C        SUBROUTINE DFTRAN INCLUDES A COMPULSORY CHECK FOR END OF DATA,
         C        PLUS OPTIONS ON CHECKS FOR MISSING DATA AND TRANSFORMATION OF
         C        DATA.  DIMENSIONED FOR A MAXIMUM OF 99 MODELS AND 100 VARIABLES
         C
         C     SUBPROGRAMS REQUIRED:
         C        DFTRAN, DFPRBF, DFPRNT, DFCRLB, DFREGR
         C
         C     ..............................................................
ISN 0002       DIMENSION TITLE(20), FMT(80)
ISN 0003       COMMON S(100),X(100),SS(100,100),STDWT(100),WTS(101),RSQ(99    )
ISN 0004       NVMAX=100
ISN 0005     1 READ(5,2,END=17) TITLE
ISN 0006     2 FORMAT(20A4)
ISN 0007       WRITE(6,5)TITLE
ISN 0008     5 FORMAT(1H1,30X,20A4)
ISN 0009       READ(5,4)NFMT,NVARIN,NVART,IPUNCH
ISN 0010     4 FORMAT(4I5)
ISN 0011       WRITE(6,6)NFMT,NVARIN,NVART,IPUNCH
ISN 0012     6 FORMAT(31H0  NFMT  NUMBER OF FORMAT CARDS,12X,I3/34H0NVARIN  NUMBE
              1R OF VARIABLES INPUT,9X,I3/40H0 NVART  NUMBER OF VARIABLES TRANSFO
              2RMED,3X,I3/29H0IPUNCH  PUNCH IDENTIFICATION,14X,I3)
ISN 0013       LFMT=20*NFMT
ISN 0014       READ(5,2)(FMT(I),I=1,LFMT)
ISN 0015       WRITE(6,7)(FMT(I),I=1,LFMT)
ISN 0016     7 FORMAT(7H0FORMAT,3X,20A4/(10X,20A4))
         C     INITIALIZE
ISN 0017       RSQ(99)=0.
ISN 0018       NPER=0
ISN 0019       DO8J=1,NVART
ISN 0020       S(J)=0.
ISN 0021       DO 8 K=J,NVART
ISN 0022     8 SS(J,K)=0.
         C     READ A DATA RECORD AND SET NEW VARIABLES, IF ANY, TO ZERO.
ISN 0023     9 READ(5,FMT)(X(J),J=1,NVARIN)
ISN 0024       IF(NVARIN.GE.NVART) GO TO 11
ISN 0026       N1=NVARIN+1
ISN 0027       DO 10 J=N1,NVART
ISN 0028    10 X(J)=0.
         C     DATA TRANSFORMATION
ISN 0029    11 CALLDFTRAN(I,X)
         C     FLAG 'I' IS RETURNED. IF NEGATIVE, GO TO NEXT CARD. IF ZERO,
         C       PROCESS THIS CARD. IF POSITIVE, END OF CARDS IS ASSUMED.
ISN 0030       IF(I)9,12,14
         C     COMPUTE SUMS AND SUMS OF CROSS PRODUCTS.
ISN 0031    12 NPER=NPER+1
ISN 0032       DO13J=1,NVART
ISN 0033       S(J)=S(J)+X(J)
ISN 0034       DO 13 K=J,NVART
ISN 0035    13 SS(J,K)=SS(J,K)+X(J)*X(K)
ISN 0036       GO TO 9
         C     COMPUTE MEANS, STANDARD DEVIATIONS, AND CORRELATIONS.
ISN 0037    14 CALL DFCRLB(NPER,NVART,NVMAX,1,NVMAX+1,2*NVMAX+1)
ISN 0038       IF(IPUNCH.EQ.0)GOTO16
ISN 0040       WRITE(7) (S(J),J=1,NVART)
```

PAGE C03

```
ISN 0041              WRITE(7)(X(J),J=1,NVART)
ISN 0042              DO 15 J=1,NVART
ISN 0043          15  WRITE(7)(SS(J,K),K=1,NVART)
ISN 0044         16   NV2=(NVMAX+1)**2
                C     DO REGRESSION ANALYSIS BY PERSUB'S REGRED (MODIFIED).
ISN 0045              CALL CFREGR(NPER,1,NVMAX+1,2*NVMAX+1,NV2,NV2+NVMAX,NV2+2*NVMAX+1,
                     1 NVMAX)
ISN CC46              GC TC 1
ISN 0C47           17 STOP
ISN 0C48              END

****** END CF COMPILATICN ******
```

```
LEVEL 13  (23 MAY 67)                    OS/360  FORTRAN H                          DATE  68.106/19.26.21

        COMPILER OPTIONS - NAME=  MR04,OPT=02,LINECNT=59,SOURCE,EBCDIC,NOLIST,NODECK,LOAD,NOMAP,NOEDIT,ID

  ISN 0002            FUNCTION DFPRBF (DN, DD, F)
                C     RETURNS PROBABILITY OF F WITH DN AND DD DEGREES OF FREEDOM.
                C     REQUIRES DFPGLF AND DFPBLF.
  ISN 0003            DIMENSION Y(6)
  ISN 0004            DFPRBF = 0.0
  ISN 0005            IF (F) 3,3,4
  ISN 0006        3   DFPRBF = 1.0
  ISN 0007            GO TO 30
  ISN 0008        4   IF (DN*DD) 30,30,5
  ISN 0009        5   Y(1) = .40546512
  ISN 0010            Y(2) = -1.2039728
  ISN 0011            Y(3) = .58778666
  ISN 0012            Y(4) = Y(2)
  ISN 0013            Y(5) = Y(1)
  ISN 0014            Y(6) = -.51082562
  ISN 0015            C = DFPGLF(DN) + DFPGLF(DD) - DFPGLF(DN+DD) - 69.693147
  ISN 0016            H = SQRT (DD/(F*DN+DD))
  ISN 0017            H=ARSIN(H)/60.
  ISN 0018            X = 0.0
  ISN 0019            DO 25 I = 1,10
  ISN 0020            IF (I-9) 15,15,10
  ISN 0021       10   Y(6) = Y(2)
  ISN 0022       15   DO 25 J = 1,6
  ISN 0023            X = X + H
  ISN 0024            XS=SIN(X)
  ISN 0025            Z = Y(J) - C + (DD-1.) *ALOG (XS) + (DN/2. - .5) *ALOG  (1.-XS*XS)
  ISN 0026            IF (Z) 25,20,20
  ISN 0027       20   DFPRBF = DFPRBF + EXP  (Z - 69.0)
  ISN 0028       25   CONTINUE
  ISN 0029            DFPRBF = DFPRBF * H
  ISN 0030       30   RETURN
  ISN 0031            END

****** END OF COMPILATION ******
```

```
LEVEL 13  (23 MAY 67)                    OS/360  FORTRAN H                          DATE  68.106/19.26.41

        COMPILER OPTIONS - NAME=  MR04,OPT=02,LINECNT=59,SOURCE,EBCDIC,NOLIST,NODECK,LOAD,NOMAP,NOEDIT,ID

             C      MEANS,STANDARD DEVIATIONS,CORRELATIONS
  ISN 0002          SUBROUTINE DFCRLB(NUM,NVAR,NVMAX,LMEAN,LSIGMA,LCORR)
             C      PORTION OF CORRLB 8 FEB/65 MODIFIED MAY/66 BY FLATHMAN, U OF A.
  ISN 0003          COMMON A(1)
             C      COMPUTE R MATRIX
  ISN 0004       80 FN=NUM
             C      COMPUTE NON DIAGONAL ELEMENTS OF R MATRIX
  ISN 0005          KM1=NVAR-1
  ISN 0006       81 DO 130 I=1,KM1
  ISN 0007       82 IP1=I+1
  ISN 0008       83 DO 130 J=IP1,NVAR
  ISN 0009       84 ISI=LCORR-1+(I-1)*NVMAX+I
  ISN 0010       85 ISJ=LCORR-1+(J-1)*NVMAX+J
  ISN 0011       86 ISIJ=LCORR-1+(J-1)*NVMAX+I
  ISN 0012       87 ISJI=LCORR-1+(I-1)*NVMAX+J
  ISN 0013       88 IMI=LMEAN-1+I
  ISN 0014       89 IMJ=LMEAN-1+J
  ISN 0015       90 DEN=SQRT((FN*A(ISI)-A(IMI)*A(IMI))*(FN*A(ISJ)-A(IMJ)*A(IMJ)))
  ISN 0016       95 IF(DEN) 110,100,110
  ISN 0017      100 A(ISIJ)=0.0
  ISN 0018      105 GO TO 130
  ISN 0019      110 A(ISIJ)=(FN*A(ISIJ)-A(IMI)*A(IMJ))/DEN
  ISN 0020      130 A(ISJI)=A(ISIJ)
             C      COMPUTE MEAN AND SIGMA
  ISN 0021      135 DO 165 I=1,NVAR
  ISN 0022      140 IM=LMEAN-1+I
  ISN 0023      145 A(IM)=A(IM)/FN
  ISN 0024      150 IS=LSIGMA-1+I
  ISN 0025      155 II=LCORR-1+(I-1)*NVMAX+I
  ISN 0026      160 A(IS)=SQRT((A(II)/FN)-A(IM)*A(IM))
             C      COMPUTE DIAGONAL ELEMENTS OF R MATRIX
  ISN 0027      165 A(II)=1.
  ISN 0028          WRITE(6,5) NUM
  ISN 0029        5 FORMAT(1HL/16HLNUMBER OF CASES,I10)
  ISN 0030          CALLDFPRNT(1,NVAR,1,A(LMEAN),8,8HMEANS    )
  ISN 0031          CALLDFPRNT(1,NVAR,1,A(LSIGMA),20,20HSTANDARD DEVIATIONS )
  ISN 0032          CALL DFPRNT(NVAR,NVAR,NVMAX,A(LCORR),12,12HCORRELATIONS)
  ISN 0033      170 RETURN
  ISN 0034          END

****** END OF COMPILATION ******
```

```
LEVEL 13  (23 MAY 67)                    OS/360  FORTRAN H                         DATE  68.106/19.26.47

        COMPILER OPTIONS - NAME=  MR04,OPT=02,LINECNT=59,SOURCE,EBCDIC,NOLIST,NODECK,LOAD,NOMAP,NOEDIT,ID

  ISN 0002            SUBROUTINE DFREGR(NPER,LMEAN,LSIGMA,LCORR,LSTDWT,LWTS,LRSQ,NVAR)
                C     REGRED FROM 8FEB/65 MODIFIED MAY/66 BY FLATHMAN. U OF A. DERS.
                C     ITERATIVE REGRESSION
  ISN 0003            DIMENSION MFLD(55),MFLDL(27),PROBID(5)
  ISN 0004            COMMON A(10500)
  ISN 0005            K6=0
  ISN 0006         31 READ (5,32) PROBID, NFLDS,STOPC,(MFLD(I),I=1,NFLDS)
  ISN 0007         32 FORMAT(5A2.I3,E8.1,28I2/21X,28I2)
  ISN 0008            IF(NFLDS) 35,35,36
  ISN 0009         35 RETURN
                C     IF NFLDS IS EVEN, GO TO 400.
  ISN 0010         36 IF(NFLDS-NFLDS/2*2)37,400,37
  ISN 0011         37 K5=NFLDS-1
  ISN 0012            IDC=MFLD(NFLDS)
  ISN 0013            WRITE(6,41)  PROBID,STOPC,IDC,(MFLD(I),I=1,K5)
  ISN 0014         41 FORMAT(1HL/5HL....5A2  ,E20.8/12H   CRITERION I5/12H  PREDICTORS I6,
                     12H -I4/(12X,I6,2H -I4))
  ISN 0015            NFLD1=NFLDS-1
  ISN 0016            DO   160 I=2,NFLD1,2
  ISN 0017            M=I/2
  ISN 0018            MFLDL(M)=MFLD(I)
  ISN 0019        160 MFLD(M)=MFLD(I-1)
                C     INITIALIZE
  ISN 0020            DO 180 I=1,NVAR
  ISN 0021            J=I+LWTS-1
  ISN 0022            A(J)=0.
  ISN 0023            J=I+LSTDWT-1
  ISN 0024        180 A(J)=0.
  ISN 0025            S=0.0
  ISN 0026            SIG2=0.0
  ISN 0027            RSQ=0.0
  ISN 0028            DEL=0.0
  ISN 0029            ITER=0
  ISN 0030            ID=1
  ISN 0031            NGRP=NFLDS/2
                C     SET FOR NEW ITERATION
  ISN 0032        220 RSQL=0.0
  ISN 0033            DO 255 I=1,NGRP
                C     ITERATE
  ISN 0034            KSTAR=MFLD(I)
  ISN 0035            KSTOP=MFLDL(I)
  ISN 0036            DO  255  J=KSTAR,KSTOP
  ISN 0037            IA=(LWTS-1)+J
  ISN 0038            IB=(LCORR-1)+((ID-1)*NVAR)+J
  ISN 0039            IC=(LCORR-1)+((IDC-1)*NVAR+J)
  ISN 0040            A(IA)=A(IA)+(DEL*A(IB))
  ISN 0041            DEN=S-(A(IA)*A(IC))
  ISN 0042            IF(DEN)245,240,245
  ISN 0043        240 DELT= A(IC)
  ISN 0044            STEST=DELT*DELT
  ISN 0045            SIG2T=STEST
  ISN 0046            RSQT=STEST
  ISN 0047            GO TO 249
  ISN 0048        245 DELT=((SIG2*A(IC))-(S*A(IA)))/DEN
  ISN 0049            STEST=S+ DELT*A(IC))
  ISN 0050            SIG2T=SIG2+(2.0*A(IA)*DELT)+(DELT*DELT)
  ISN 0051            RSQT=(STEST*STEST)/SIG2T
  ISN 0052        249 IF(RSQL-RSQT)250,255,255
```

```
ISN 0053      250  SLAR=STEST
ISN 0054           SIG2L=SIG2T
ISN 0055           RSQL=RSQT
ISN 0056           DELTL=DELT
ISN 0057           IDLAR=J
ISN 0058      255  CONTINUE
ISN 0059           IF(RSQL-RSQ-STOPC)268,256,256
ISN 0060      256  S=SLAR
ISN 0061           SIG2=SIG2L
ISN 0062           RSQ=RSQL
ISN 0063           DEL=DELTL
ISN 0064           ITER=ITER+1
ISN 0065           ID= IDLAR
ISN 0066           IA = (LSTDWT-1) + ID
ISN 0067           A(IA)=A(IA)+DEL
ISN 0068           IF(RSQ-1.) 220,268,1266
ISN 0069     1266  WRITE(6,1267)
ISN 0070     1267  FORMAT(25H RSQ IS GREATER THAN ONE.)
ISN 0071           STOP
              C    TERMINATE
ISN 0072      268  SDS2=S/SIG2
ISN 0073           WRITE(6,265) RSQL,ITER
ISN 0074      265  FORMAT(1HJ,4X,5HRSQ =F11.8,40X,I5)
ISN 0075           DO  274   I=1,NGRP
ISN 0076           KSTAR=MFLD(I)
ISN 0077           KSTOP=MFLDL(I)
ISN 0078           DO  274 J= KSTAR,KSTOP
ISN 0079           IA=LSTDWT-1+J
ISN 0080      274  A(IA)=A(IA)*SDS2
ISN 0081           WRITE(6,278)
ISN 0082      278  FORMAT(41HK VAR. NUMBER       STD. WT.        ERROR/)
ISN 0083           DO 295 I=1,NGRP
ISN 0084           KSTAR=MFLD(I)
ISN 0085           KSTOP=MFLDL(I)
ISN 0086           DO 295  J=KSTAR.KSTOP
ISN 0087           IA=LWTS-1+J
ISN 0088           A(IA)=0.0
ISN 0089           DO  291  IL=1,NGRP
ISN 0090           LSTAR=MFLD(IL)
ISN 0091           LSTOP=MFLDL(IL)
ISN 0092           DO  291  L=LSTAR.LSTOP
ISN 0093           IB=LSTDWT-1+L
ISN 0094           IC=LCORR-1+J+((L-1)*NVAR)
ISN 0095      291  A(IA)=A(IA)+(A(IB)*A(IC))
ISN 0096           IC=LCORR-1+J+((IDC-1)*NVAR)
ISN 0097           A(IA)=A(IA)-A(IC)
ISN 0098           IB=LSTDWT-1+J
ISN 0099      295  WRITE(6,296)  J,A(IB),A(IA)
ISN 0100      296  FORMAT(1H I9,F18.8,F15.8)
              C    COMPUTE REGRESSION EQUATION
ISN 0101           WRITE(6,299)
ISN 0102      299  FORMAT(  27HK VAR. NUMBER        WEIGHT/)
ISN 0103           FK1=0.0
ISN 0104           DO  315 I=1,NGRP
ISN 0105            KSTAR=MFLD(I)
ISN 0106           KSTOP=MFLDL(I)
ISN 0107           DO  315  J=KSTAR,KSTOP
ISN 0108           IA=LSIGMA-1+J
ISN 0109           IB=LSTDWT-1+J
```

PAGE 003

```
ISN 0110          IC=LSIGMA-1+IDC
ISN 0111          ID=LMEAN-1+J
ISN 0112          IE=LWTS-1+J
ISN 0113          IF(A(IA))313,311,313
ISN 0114      311 A(IE)=0.0
ISN 0115          GO TO 315
ISN 0116      313 A(IE)=A(IB)*(A(IC)/A(IA))
ISN 0117          FK1=FK1+(A(IB)*(A(ID)/A(IA)))
ISN 0118     315  WRITE(6,316)  J,A(IE)
ISN 0119      316 FORMAT(1H I9,F18.8)
ISN 0120          ID = LMEAN-1+ IDC
ISN 0121          REGCO=A(ID)-(A(IC)*FK1)
ISN 0122          WRITE(6,330) REGCO
ISN 0123      330 FORMAT ( 10HJCONSTANT=F18.8)
ISN 0124          LAPN=LWTS+NVAR
ISN 0125          A(LAPN)=REGCO
ISN 0126          K5=LRSQ+K6
ISN 0127          A(K5)=RSQL
ISN 0128          K6=K6+1
ISN 0129          GO TO 31
           C          COMPUTE F RATIO
ISN 0130      400 DF1=MFLD(3)*100+MFLD(4)
ISN 0131          DF2=MFLD(5)*100+MFLD(6)
ISN 0132          K8=MFLD(1)-1+LRSQ
ISN 0133          K9=MFLD(2)-1+LRSQ
ISN 0134          IF(NFLDS.NE.8) GO TO 601
           C      IF NFLDS IS 8, DF1 AND DF2 ARE READ AS THE DEGREES OF FREEDOM OF
           C          THE FULL AND RESTRICTED MODELS, RESPECTIVELY.
ISN 0136          FNPER=NPER
ISN 0137          DF1=DF1-DF2
ISN 0138          DF2=FNPER-DF1-DF2
ISN 0139      601 F=((A(K8)-A(K9))/DF1) / ((1.0-A(K8))/DF2)
ISN 0140          P=DFPRBF(DF1,DF2,F)
ISN 0141          WRITE(6,406) PROBID,F,DF1,DF2,MFLD(1),A(K8),MFLD(2),A(K9),P
ISN 0142      406 FORMAT(1HK,5A2 ,4H F = ,     F10.4,3X,12H D.F. NUM. =F5.0,12H D.F.
                 1DEN. =F8.0,3X,I4,F8.5,I3,F8.5,2X,6HPROB =,F8.5)
ISN 0143          GO TO 31
ISN 0144          END

****** END OF COMPILATION ******
```

```
LEVEL 13  (23 MAY 67)                     OS/360  FORTRAN H                          DATE  68.106/19.26.35

        COMPILER OPTIONS - NAME=  MR04,OPT=02,LINECNT=59,SOURCE,EBCDIC,NOLIST,NODECK,LOAD,NOMAP,NOEDIT,ID

  ISN 0002               SUBROUTINE DFPRNT(NR,NC,NRMAX,A,NUMHOL,TITLE)
                  C      PRINTS A TWO-DIMENSIONAL ARRAY 'A' WITH NR ROWS AND NC COLUMNS.
                  C      NRMAX IS THE MAXIMUM NUMBER OF ROWS DIMENSIONED FOR @A@.
                  C      FOR A LINEAR ARRAY, SET NR=NRMAX=1
                  C      'TITLE' MUST BE A MULTIPLE OF 4
  ISN 0003               DIMENSION A(1),TITLE(20)
  ISN 0004               N=(NUMHOL+3)/4
  ISN 0005               WRITE(6,4) (TITLE(J),J=1,N)
  ISN 0006          4    FORMAT(1HK/1HK,20A4)
  ISN 0007               N=(NC-1)/10+1
  ISN 0008               DO 1 K=1,N
  ISN 0009               JA=K*10-9
  ISN 0010               JB=K*10-K/N*(K*10-NC)
  ISN 0011               WRITE(6,2) (J,J=JA,JB)
  ISN 0012          2    FORMAT(1HK,8X,10I12)
  ISN 0013               WRITE (6,3)
  ISN 0014          3    FORMAT(1HJ)
  ISN 0015               JA=(JA-1)*NRMAX
  ISN 0016               JB=(JB-1)*NRMAX
  ISN 0017               DO 1 I=1,NR
  ISN 0018               JA=JA+1
  ISN 0019               JB=JB+1
  ISN 0020          1    WRITE(6,5) I,(A(J),J=JA,JB,NRMAX)
  ISN 0021          5    FORMAT(I5,5X,10F12.4)
  ISN 0022               RETURN
  ISN 0023               END

****** END OF COMPILATION ******


LEVEL 13  (23 MAY 67)                     OS/360  FORTRAN H                          DATE  68.106/19.26.31

        COMPILER OPTIONS - NAME=  MR04,OPT=02,LINECNT=59,SOURCE,EBCDIC,NOLIST,NODECK,LOAD,NOMAP,NOEDIT,ID

  ISN 0002               FUNCTION DFPBLF (X)
  ISN 0003               DFPBLF = 0.0
  ISN 0004               IF (X-2.0) 10,5,5
  ISN 0005          5    Z = 1.0 / (X * X)
  ISN 0006               DFPBLF = (X+.5)*ALOG(X) - X+.91893853 + (.083333333-Z * (.0027778
                        1 - Z * (.00079365 - Z * (.00059524 - Z * .00084175)))) / X
  ISN 0007         10    RETURN
  ISN 0008               END

****** END OF COMPILATION ******
```

```
LEVEL 13  (23 MAY 67)                    OS/360   FORTRAN H                     DATE  68.106/19.26.16

        COMPILER OPTIONS - NAME=  MR04,OPT=02,LINECNT=59,SOURCE,EBCDIC,NOLIST,NODECK,LOAD,NOMAP,NOEDIT,ID

  ISN 0002             SUBROUTINE DFTRAN(I,X)
                 C     DATA TRANSFORMATION
                 C     FLAG 'I' IS RETURNED. IF NEGATIVE, GO TO NEXT CARD.  IF ZERO,
                 C        PROCESS THIS CARD.  IF POSITIVE, END OF CARDS IS ASSUMED.
  ISN 0003             DIMENSION X(1)
  ISN 0004             I=1
                 C     COMPULSORY CHECK FOR END-OF-DATA.  IF SO, RETURN.
  ISN 0005             IF(X(1).LT.0.) RETURN
  ISN 0007             I=-1
                 C     OPTIONAL CHECK FOR MISSING DATA.  IF SO, RETURN.
                 C     WHEN CHECKING FOR MISSING DATA IT IS ASSUMED THAT SCORES OF
                 C        0 DO NOT EXIST.
                 C*****TO MAKE THE OPTIONAL CHECK,DEFINE 'N' AS THE NUMBER OF VARIABLES
                 C        AND REMOVE THE 'C' FROM COLUMN ONE OF THE FOLLOWING 4 CARDS.
                 C     N=
                 C     DO 1 J=1,N
                 C     IF(X(J).EQ.0.) RETURN
                 C   1 CONTINUE
                 C     INSERT DATA TRANSFORMATION CARDS BETWEEN HERE AND 'I=0' CARD.
  ISN 0008              IF(X(2).EQ.1) X(3)=1.
  ISN 0010             IF(X(2).EQ.2) X(4)=1.
  ISN 0012             IF(X(2).EQ.3) X(5)=1.
  ISN 0014             IF(X(2).EQ.4) X(6)=1.
  ISN 0016             X(7)=X(1)*X(3)
  ISN 0017             X(8)=X(1)*X(4)
  ISN 0018             X(9)=X(1)*X(5)
  ISN 0019              X(10)=X(1)*X(6)
  ISN 0020             I=0
  ISN 0021             RETURN
  ISN 0022             END

****** END OF COMPILATION ******


LEVEL 13  (23 MAY 67)                    OS/360   FORTRAN H                     DATE  68.106/19.26.26

        COMPILER OPTIONS - NAME=  MR04,OPT=02,LINECNT=59,SOURCE,EBCDIC,NOLIST,NODECK,LOAD,NOMAP,NOEDIT,ID

  ISN 0002             FUNCTION DFPGLF(X)
  ISN 0003             DFPGLF = .57236493
  ISN 0004             IF (X - 1.0) 15,15,5
  ISN 0005       5     L = X
  ISN 0006             FL = L/2 * 2
  ISN 0007             IF(X-FL) 20,20,10
  ISN 0008      10     DFPGLF = DFPGLF - (X-2.)*.69314718 + DFPBLF(X-2.)-DFPBLF(X/2.-1.5)
  ISN 0009      15     RETURN
  ISN 0010      20     DFPGLF = DFPBLF (X/2.0 - 1.0)
  ISN 0011             RETURN
  ISN 0012             END

****** END OF COMPILATION ******
```

References

Bender, J. A.; F. J. Kelly; J. K. Pierson; and H. M. Kaplan, "Analysis of the Comparative Advantages of Unlike Exercises in Relation to Prior Individual Strength Level," Research Quarterly (in press).

Blommers, P., and E. F. Lindquist, Elementary Statistical Methods, Boston: Houghton Mifflin, 1960.

Bottenberg, R. A., and J. H. Ward jr., Applied Multiple Linear Regression, U. S. Department of Commerce Office of Technical Services, AD413128, 1963.

Box, G. E. P., "Some Theorems on Quadratic Forms Applied in the Study of Analysis of Variance Problems, I. Effect of the Inequality of Variance in the One-way Classification," Annals Mathematical Statistics, 25, 1954, 290-302.

Castaneda, A.; D. S. Palermo; and B. McCandless, "Complex Learning and Performance as a Function of Anxiety in Children and Task Difficulty," Child Development, 27, 1956, 328-332.

Cattell, R., The Scientific Analysis of Personality, Chicago: Aldine Publishing Company, 1966.

Cellura, R., The application of psychological theory in educational settings: An overview. A.E.R.A. convention, Chicago, Illinois, 1968.

Cochran, W. G., Sampling Techniques, New York: John Wiley & Sons, 1953.

Denny, J. P., "Effects of Anxiety and Intelligence on Concept Formation," Journal of Experimental Psychology, 72, 1966, 596-602.

Edwards, A. L., Statistical Methods for the Behavioral Sciences, New York: Holt, Rinehart, and Winston, Inc., 1964.

Foster, G., Some multiple linear regression models for the evaluation of school programs. A.E.R.A. convention, Chicago, Illinois, 1968.

Gagné, R. M., *The Conditions of Learning*, New York: Holt, Rinehart, and Winston, Inc., 1965.

Guilford, J. P., "Intelligence: 1965 Model," *American Psychologist*, *21*, 1966, 20-26.

Guilford, J. P., *Fundamental Statistics in Psychology and Education* (fourth ed.), New York: McGraw-Hill, 1965.

Guilford, J. P., "Three Faces of the Intellect," *American Psychologist*, *14*, 1959, 469-479.

Guilford, J. P., "Creativity," *American Psychologist*, *9*, 1950.

Harris, C. W. (Ed.), *Problems in Measuring Change*, Madison, Wisconsin: The University of Wisconsin, 1963.

Hays, W. L., *Statistics for Psychologists*, New York: Holt, Rinehart, and Winston, Inc., 1963.

Horst, P., "Relations among M Sets of Measures," *Psychometrika*, *26*, 1961, 129-149.

Jennings, E., "Matrix Formulas for Part and Partial Correlation," *Psychometrika*, *30*, 1965, 353-356.

Katahn, M., "Interaction of Anxiety and Ability in Complex Learning Situations," *Journal of Personality and Social Psychology*, *3*, 1966, 475-478.

Lindquist, E. F., *Design and Analysis of Experiments in Psychology and Education*, Boston: Houghton Mifflin, 1953.

McClelland, D. C., *The Achievement Motive*, New York: Appleton-Century-Crofts, 1953.

McGuire, C., "The Textown Study of Adolescence," *Texas Journal of Science*, *8*, 1956, 264-274.

McGuire, C., "The Prediction of Talented Behavior in the Junior High School," *Invitational Conference on Testing Problems*, Princeton, New Jersey: Educational Testing Service, 1960, 46-73.

Norton, D. W., An empirical investigation of some effects of non-normality and heterogeneity on the F-distribution. Unpublished doctoral dissertation, State University of Iowa, 1952.

Osgood, C. E.; G. J. Suci; and P. H. Tannenbaum, The Measurement of Meaning, Urbana, Illinois: University of Illinois Press, 1957.

Phillips, B. N.; R. Adams; and K. A. McNeil, Academic motivation and school anxiety as concomitants of change in measured intelligence. Paper presented at S.W.P.A., 1967.

Runyon, R. P., and A. Haber, Fundamentals of Behavioral Statistics, Reading, Massachusetts: Addison-Wesley, 1967.

Sarason, I. G., "Test Anxiety, General Anxiety, and Intellectual Performance," Journal of Consulting Psychology, 21, 1957, 485-490.

Sarason, I. G., and E. G. Palola, "The Relationship of Test and General Anxiety, Difficulty of Task, and Experimental Instructions to Performance," Journal of Experimental Psychology, 59, 1960, 185-191.

Snedecor, G. W., Statistical Methods, Ames, Iowa: The Collegiate Press, Inc., 1946.

Spence, J. T.; B. J. Underwood; C. P. Duncan; and J. W. Cotton, Elementary Statistics (second ed.), New York: Appleton-Century-Crofts, 1968.

Suppes, P.; L. Hyman; and M. Jerman, "Linear Structural Models for Response and Latency Performance in Arithmetic," Technical Report 100, Stanford, California: Institute for Mathematical Studies in the Social Sciences, 1966.

Taylor, J., "A Personality Scale of Manifest Anxiety," Journal of Abnormal and Social Psychology, 48, 1953, 285-290.

Veldman, D. J., FORTRAN Programming for the Behavioral Sciences, New York: Holt, Rinehart, and Winston, Inc., 1967.

Whiteside, R., Dimensions of teacher evaluation of academic achievement. Unpublished doctoral dissertation, The University of Texas, Austin, Texas, 1964.

Whiting, J. W. M., and I. L. Child, Child Training and Personality, New Haven: Yale University Press, 1953.

Winer, B. J., Statistical Principles in Experimental Design, New York: McGraw-Hill, 1962.

Index